MECHANICAL DRAWING PRO

FOR

HIGH SCHOOLS, NORMAL SCHOOLS AND VOCATIONAL S

BY EDWARD BERG AND EMIL F. KRONQUIST

Instructions in Mechanical Drawing
Washington High School
Milwaukee, Wis.

THE MANUAL ARTS PRESS
PEORIA, ILLINOIS

PREFACE

The aim of the authors is to present to the student and the teacher a collection
problems embodying the fundamental principles and examples of practical mecha
arranged to cover two years of school work. The problems are grouped into first s
semester, and third and fourth semester work, and arranged with the view in mind
proper amount of principles and application during a given period of time. With
felt that the student who leaves school at any time during the first two years or
two years, will have received the proportional amount of mechanical drawing which
needs best in the practical walks of life.

Each problem is given in the form of a specification sheet and a lay-out she
the student is to make the completed drawing. The specification sheet gives a st
problem and is frequently supplemented with text matter which bears upon and
thing to be taught in the problem, or gives some relative information. The lay-out
the method of procedure, and also furnishes an object lesson in the form of care
work which is always before the pupil. With this method, class instruction is redu
mum, giving the teacher more time for individual instruction, and the pupil the
working things out by himself. The proper use of the specification, lay-out, and r
will do much, it is believed, towards creating self-reliance and personal effort ar
the part of the student.

The required problems are fully supplemented, making the course flexible, ar
many needs. The reference sheets are made use of thruout the book, and will be
able feature. The extra plates may be used to extend the work over a period
two years.

OUTLINE OF COURSE BY SEMESTERS.

FIRST SEMESTER

Plates 1, 2, 3, 8, 12, and 15 are required inked or traced. As many other plates as t should be inked or traced. If possible, a blueprint should be made from tracing of plate 1

The supplementary problems are intended for the student that works ahead of the class. stituted for the corresponding required problems or be given as additional problems. They test plates.

SECOND SEMESTER

Plates 16, 27, 28, 29, 30, 31, and 32 are required inked or traced. As many other
permit should be inked or traced. If possible, a blueprint should be made from tracing

The supplementary problems are intended for the student that works ahead of the cl
stituted for the corresponding required problems or be given as additional problems. The
test plates.

THIRD AND FOURTH SEMESTERS

Plates 42 to 49 inclusive, should be inked or traced. If possible, blueprints should be of plates 45 and 49.

The supplementary problems are intended for the student that works ahead of the class. stituted for the corresponding required problems or be given as additional problems. They i test plates.

REFERENCE SHEETS

EXTRA PLATES

LIST OF INSTRUMENTS AND MATERIALS.

1. Set of drawing instruments, including at least one ruling pen and a changeable lead point and pen point.
2. Drawing board.
3. T-square.
4. 45° and 30°—60° triangles.
5. Irregular curve.
6. 12-inch architect's triangular scale.
7. One 6H and one 3H pencil.
8. Bottle of drawing ink.
9. 1 doz. flat head thumb-tacks.
10. Penholder, assorted writing pens.
11. Pencil eraser and cleaning rubber.
12. Piece of soft cloth to keep ruling pen clean, and a few small sheets (pencil points sharp.

The drawings are made on 11″ x 15″ sheets of paper which are cut from sheet which is a standard size and measures 22″ x 30″. They are trimmed to the drawing is completed. A good grade of cream colored or white paper is 1

LAY-OUT SHEET

① _With scale and pencil used as indicated, locate points for horizontal lines._

② _Through the points d of indefinite length. Us T-square._

③ _With scale and pencil used as indicated, locate points for vertical lines._

④ _With pencil, T-square draw vertical lines tom towards the top_

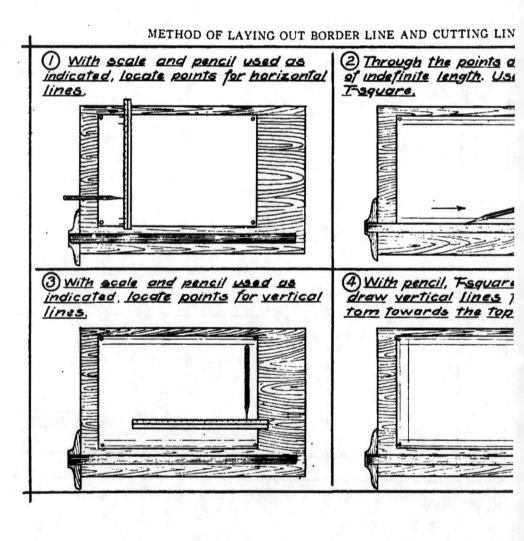

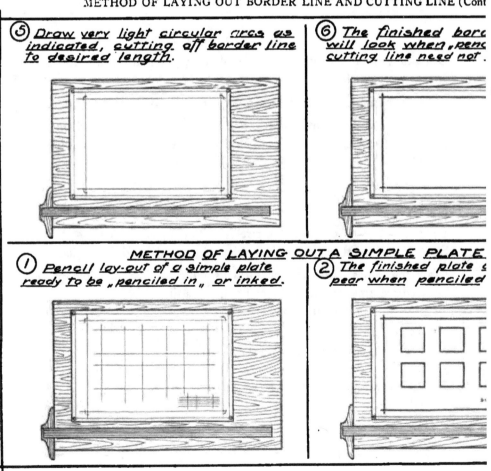

⑤ *Draw very light circular arcs as indicated, cutting off border line to desired length.*

⑥ *The finished bord will look when penc cutting line need not .*

METHOD OF LAYING OUT A SIMPLE PLATE

① *Pencil lay-out of a simple plate ready to be „penciled in„ or inked.*

② *The finished plate pear when penciled*

Specification—Plate 1.

Draw horizontal and vertical lines with T-square and triangle to form eight : the lay-out sheet on the opposite page.

Draw diagonals in squares 1, 2, 3, 4, 5, 6, and space off half inches on the'

Sq. 1.   Draw horizontal and vertical lines through half inch marks.

Sq. 2.   "   45° lines (to a horizontal) through half inch marks in bot

Sq. 3.   "   30° "   " "   "   "   "   "   "   "   "

Sq. 4.   "   60° "   " "   "   "   "   "   "   "   "

Sq. 5.   "   15° "   " "   "   "   "   "   "   "   "

Sq. 6.   "   75° "   "." "   "   "   "   "   "   "   "

Sq. 7.   "   five concentric circles.

Sq. 8.   "   three eccentric circles 2½", 1¾" and 1" in diameter.

Draw four lines used in mechanical drawing; visible, invisible, projection a: Draw three arcs having 9, 8, and 7 inch radii. Omit all dimensions and figures.

1. What is an angle?
2. How many degrees are there in a circle?
3. How many degrees are there in a right angle?
4. What is meant by concentric and eccentric circles?

PLATE 1

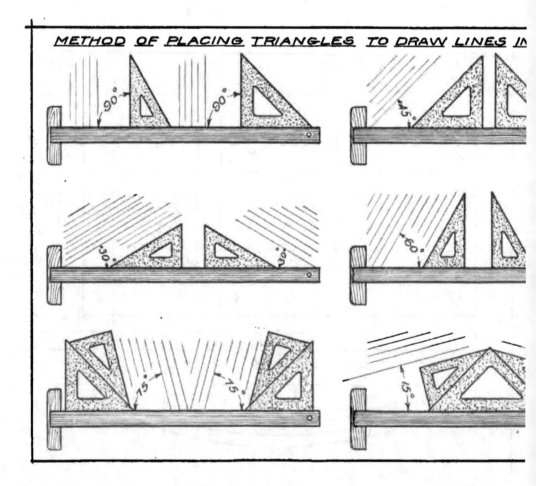

PLATE 1 (Completed)

Robert

Specification—Plate 2.

This plate is to acquaint the student with the form and method of exe
capital letters and to lay the foundation for the lettering on the plates that are

The letters are arranged in three groups: the first group comprising 1
horizontal strokes; the second group including slant or inclined strokes; the
curved strokes. A number of words are given with each group. Enough 1
tered to fill out the space. Shorter words such as IF, IT, TILL, WE, AX
used or original words may be substituted by the student.

In laying out the sheet draw first the horizontal guide lines. The bow d
the height of the letters and used to good advantage. Then draw the vertic
letters in the groups, spacing them evenly as indicated in the lay-out sheet. T
should be close to each other and the words well separated. Draw a few vert
ing them at random, to help form the letters in the words and also the figur
each stroke should be followed carefully and should always be used in maki

The title should always be symmetrical with reference to a vertical cent
located as indicated on the lay-out sheet. Start with the longest line and let
will insure its being placed correctly and will help in making it symmetrical.

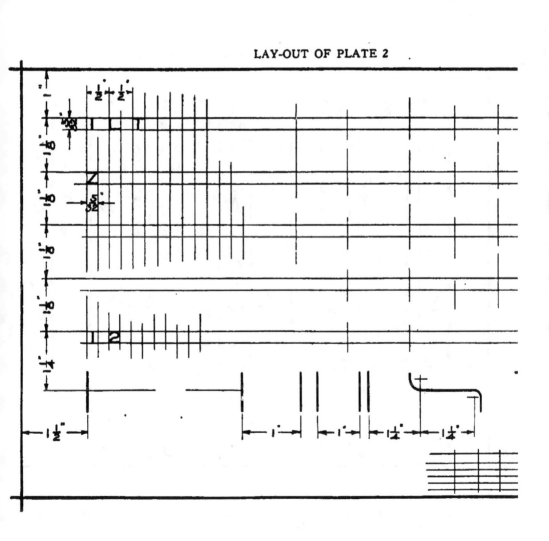

Location dimensions are always laid out to full size, but omitted in the finished drawing.

PLATE 2 (Completed)

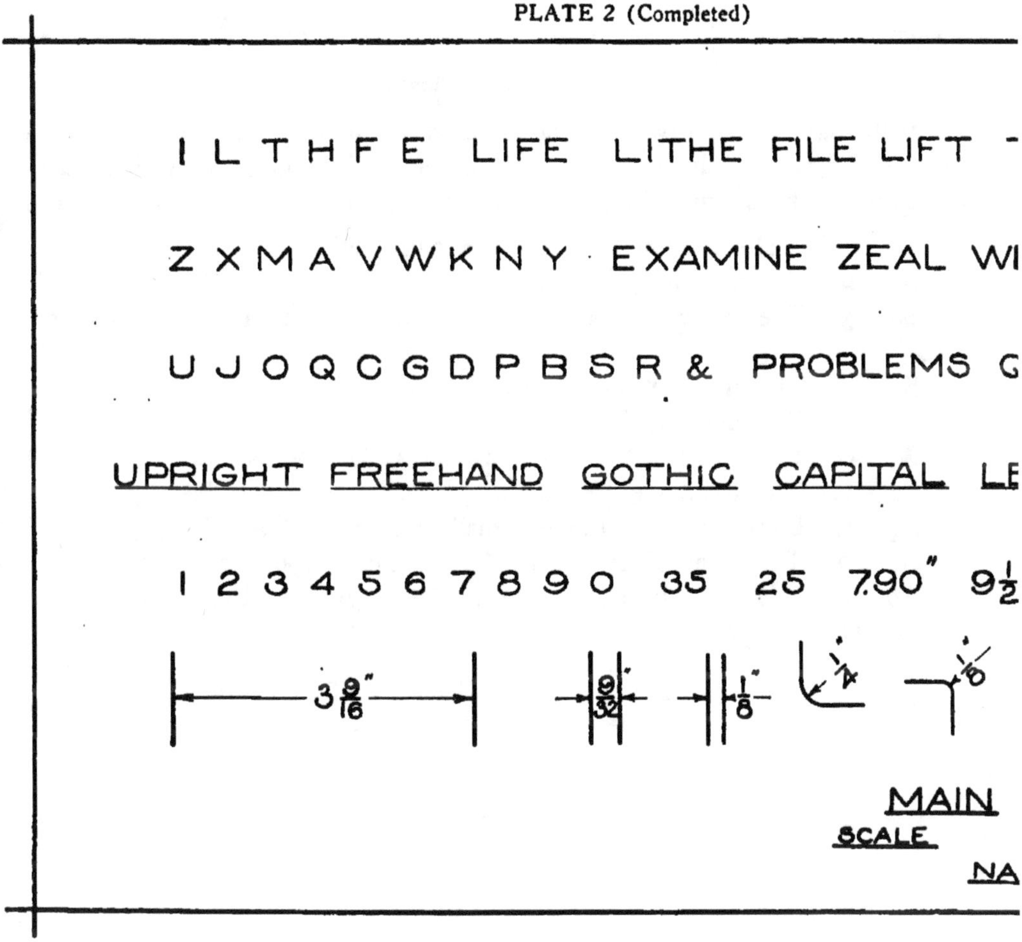

I L T H F E   LIFE   LITHE   FILE   LIFT ⌐

Z X M A V W K N Y · EXAMINE ZEAL WI

U J O Q C G D P B S R & PROBLEMS C

UPRIGHT FREEHAND GOTHIC CAPITAL LE

1 2 3 4 5 6 7 8 9 0   35   25   7.90″ 9½

MAIN

SCALE

NA

Specification—Plate 3.

With light lines draw 11 squares as shown in the plate on the opposite contains a figure involving a geometric problem, the solution of which must the figure can be drawn. Study carefully the geometric problem referred to i

Sq. 1.   Make the line 2″ long. See page 188, Prob. 1.

Sq. 2.   See page 188, Prob. 2.

Sq. 3.   The vertex of the angle is located as shown. The angle is ᴜ

Prob. 3.

Sq. 4.   The line is 2″ long and is to be divided into 5 parts. See page

Sq. 5.   See page 189, Prob. 5.

Sq. 6.   Large arcs, 1″ radius; small arcs, ½″ radius. See page 189,]

Sq. 7.   The arcs have a radius of ½″. See page 190, Prob. 7.

Sq. 8.   Large arc, ¾″ radius; small arcs ⅜″ radius. See page 190, F

Sq. 9.   Diameter of circle, 2½″. See page 190, Prob. 9.

Sq. 10.   Diameter of circle, 2⅛″. See page 191, Prob. 10.

Sq. 11.   See page 191, Prob. 11.

Title:— GEOMETRIC PROBLEMS

SCALE   DATE

NAME

PLATE 3

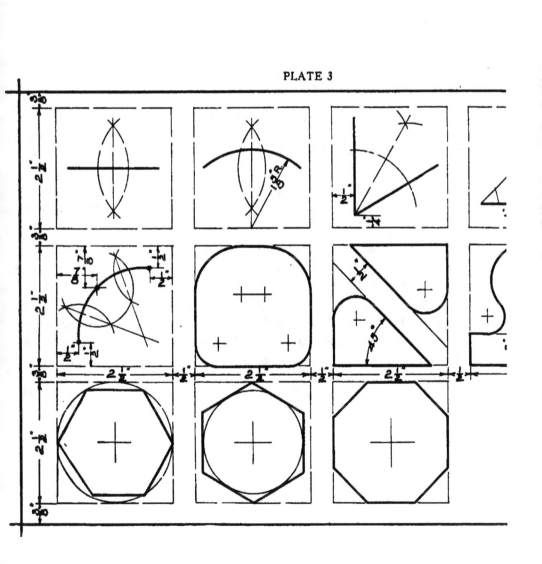

WORKING DRAWINGS.

A working drawing of an object is a group of completely dimensioned
so arranged and drawn that it will give all the information necessary to mak

A picture drawing of an object is a single view of the object represented
eye when viewed or looked at from a stationary point. It shows only those
edges, that can be seen from one point and does not show them in their tru
or relative size.

In a working drawing the object is viewed from many points—as many
show all the edges, surfaces, etc., in their true shape and size. In a picture di
view of the object, in a working drawing we have two or more views, someti

It will be noticed in the working drawings on the opposite page that the
edges of an object are shown as well as the visible edges. They are repres
broken lines which indicate that they are hidden from view by some other part

1. How does a working drawing differ from a picture drawing of an object?
2. How many views should be shown in a working drawing of an object?
3. How are invisible or hidden edges of an object indicated in a working drawi

WORKING DRAWINGS

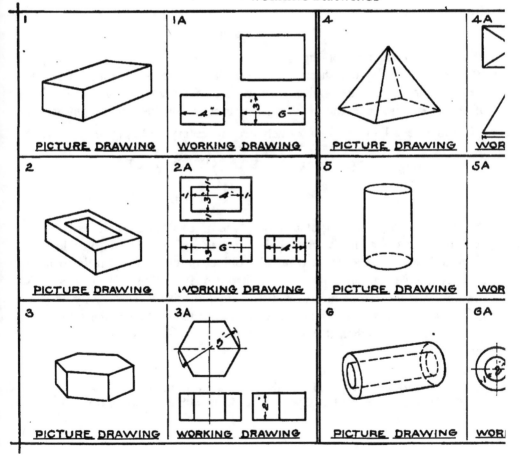

1	**1A**	**4**	**4A**
PICTURE DRAWING	WORKING DRAWING	PICTURE DRAWING	WOR
2	**2A**	**5**	**5A**
PICTURE DRAWING	WORKING DRAWING	PICTURE DRAWING	WOR
3	**3A**	**6**	**6A**
PICTURE DRAWING	WORKING DRAWING	PICTURE DRAWING	WOR

Specification—Plate 4.

Make a complete working drawing of the Lap-joint Piece, showing the side views. The lay-out sheet on the opposite page gives the location of views view complete. The top and side views are partially drawn and are to be co of the front view and the picture drawing shown in the upper right-hand co

Indicate all dimensions which are necessary and helpful to make the

Title:— LAP-JOINT PIECE

SCALE   DATE

NAME

In a working drawing of an object one of its sides or surfaces is selecte It is drawn as it appears when held squarely in front of and on a level wi view is drawn as it appears when the object is viewed squarely from al directly above and in line with the front view. The side view is drawn the object is viewed squarely from the side, and is placed directly opposite front view. When the object is viewed from the right side, the side view i of the front view; when it is viewed from the left side, the side view is pla front view.

1. Is the top view placed directly above the front view?
2. Is the side view placed directly opposite the front view?
3. Where should the left side view be placed with reference to the front view?

PLATE 4

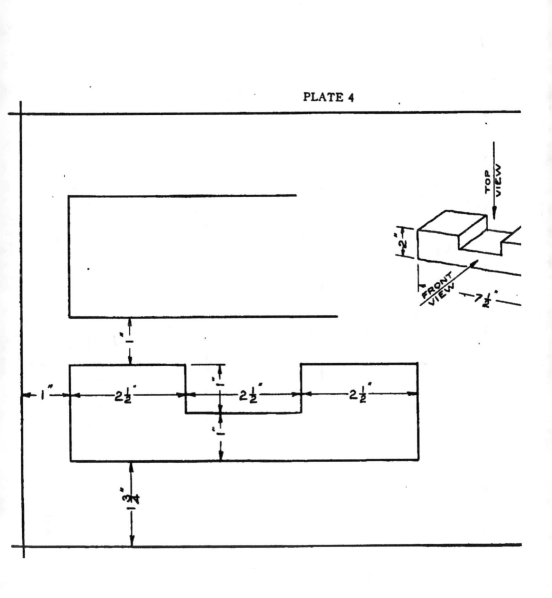

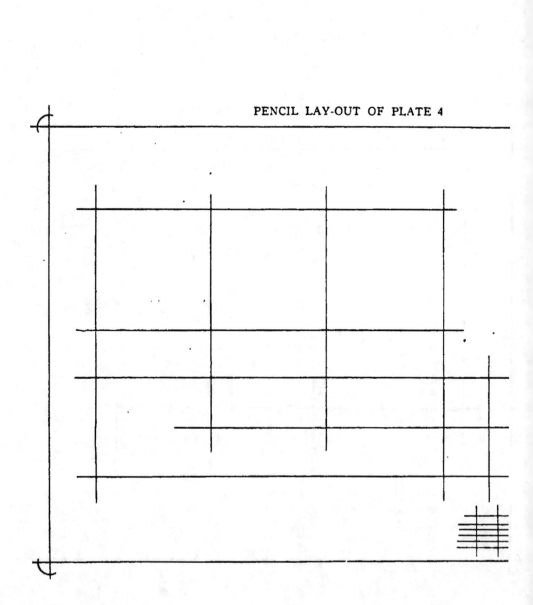

PLATE 4 (Completed)

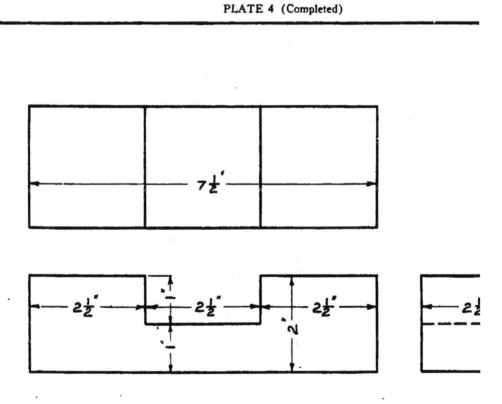

LAP JOIN
SCALE 12"=1'
Sign N

Note. It will be noticed that all dimensions are placed so as to read from the bottom

Specification—Plate 5.

Make a complete working drawing of the T-slot base, showing the fr and right-side view. The lay-out sheet gives the location of views and sl completed. The front and top views are to be drawn with the aid of the picture drawing.

Indicate all necessary dimensions.

Title :—   T-SLOT BASE

SCALE   DATE

NAME

Any side of an object can be used as the front view. It is customary the object in its natural position and to use that side which shows clearly the front view. When the front view has been determined upon, the t(must be drawn in their proper relation to it.

1. In making a working drawing can any side of an object be used as the front
2. In what position is an object generally drawn?
3. Which side of an object is preferred for the front view?

PLATE 5

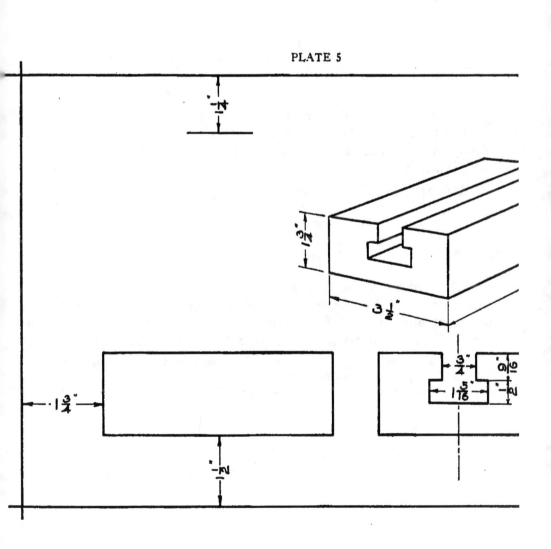

Specification—Plate 6.

Make a complete working drawing of the V-block, showing the fron
right-side view. The lay-out sheet gives the location of views and shows
pletely drawn. The top and side views are to be drawn with the aid o
the perspective sketch.

Indicate all necessary dimensions.

Title :—   V-BLOCK
           SCALE   DATE
                      NAME

The number of views in a working drawing depends upon the sha
nature of the article drawn. A good drawing must show enough viev
the shape and size of the different parts, surfaces and edges, and their
other. This may necessitate three or more views and should have at le

1. How many views should be shown in a working drawing of an object?
2. Why is one view not sufficient in a complete working drawing?
3. What is the weight, in cast iron, of the V-Block? The weight of cast iron i

PLATE 6

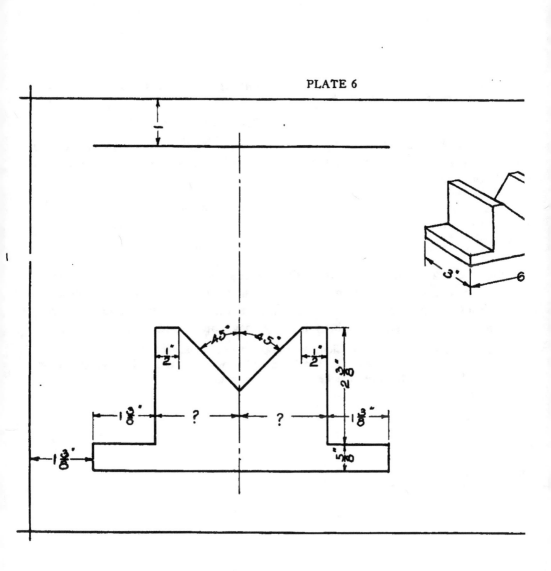

Specification—Plate 7.

With the aid of the perspective sketch, make a complete working dr
showing the front view, top view and right-side view. Include all n
Scale 12″—1′.

Title :—   BRACKET

       SCALE   DATE

           NAME

Correct and well-placed dimensions are a very important part of
They are fully as important as the lines which indicate the shape of t
should be placed carefully. The figures and arrow-heads should be well
may be read easily and their meaning may not be mistaken.

1. Why are the dimensions on a drawing very important?
2. Is the drawing of the Bracket complete with the front view and side view?
3. Could all dimensions be shown in the front view?

PLATE 7

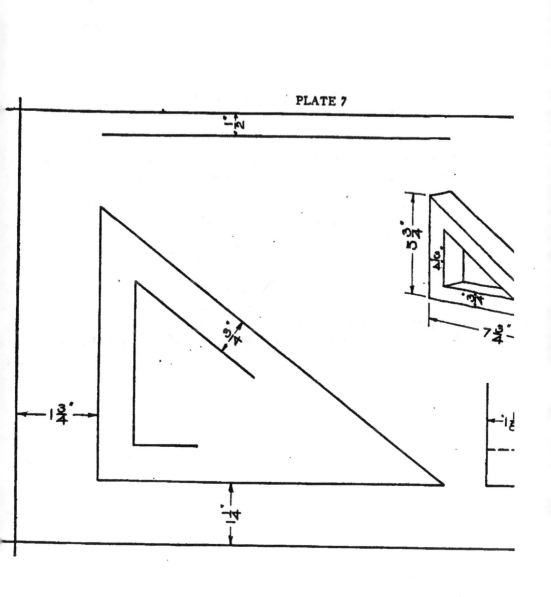

Specification—Plate 7A.

With the help of the perspective sketch, make a complete working c
showing the front view, top view and right-side view.   Indicate all ɪ
Scale 12″—1′.

Title:—   BRACE
                SCALE   DATE
                          NAME

NOTES ON DIMENSIONING.

1. Dimensions should read from the bottom and the right side of the sheet.
2. In general, dimensions should not be repeated.
3. Figures in fractions should be made large enough to be read easily.

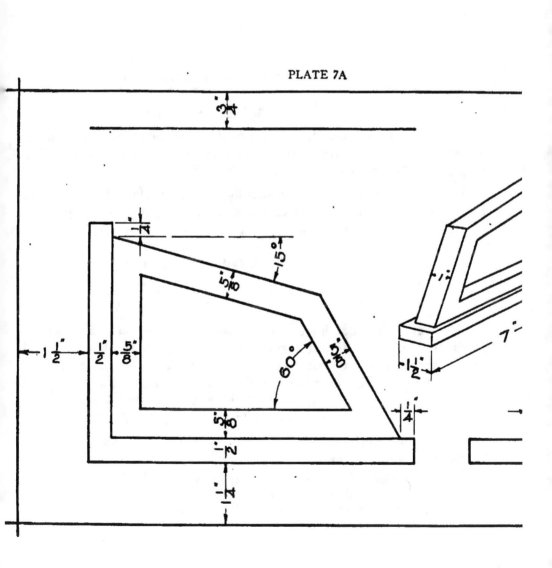

Specification—Plate 8.

Make a complete working drawing of the Ink-well Stand, showing top view, and the right-side view "in section" or as it would be seen if p drawing were removed.

The section lines should be drawn light, and should indicate the kin The Ink-well Stand is made of cast-iron, wood or glass. See page 196.

The chamfers are ¼″ x ¼″. Scale 12″=1′.

Title:—   INK-WELL STAND
SCALE   DATE
NAME

When an object has interior construction or is of such a shape tha broken lines to show hidden parts, it is advantageous to show one of t of the object removed or cut away so that the shape or construction r clearly and the dimensions placed to the best advantage. Such a view is tion" or partly in section and the surface or surfaces which have been sawed, are covered with "section lines" representing the kind of material t

Section lines are generally the last lines to be drawn and should not dimensions have been placed.

1. Why are views of an object or parts of views shown in section?
2. What are section lines and why are they used?
3. Why is the drawing of the Ink-Well Stand not complete with the front and t

PLATE 8

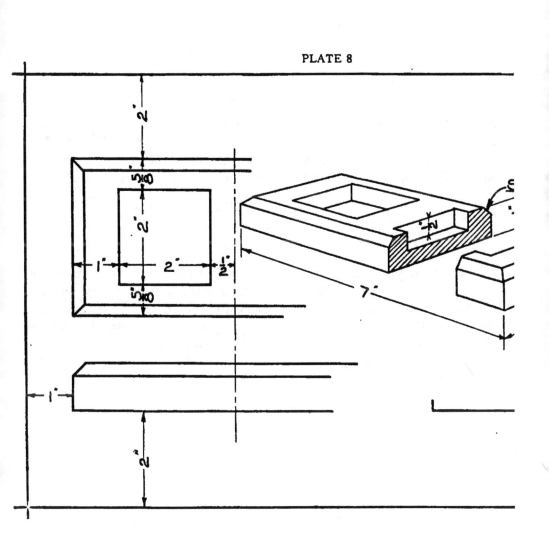

Specification—Plate 8A.

Make a complete working drawing of the Try-square, including t
view, left-side view and an isolated section of the beam. Scale 12″=1′.

The graduations, or division marks on the blade, are to be drawn
are ⅛″, ¼″, ⅜″ and ½″ long respectively.   The blade is ¹⁄₁₆″ thick.

Title:—   TRY-SQUARE
              SCALE   DATE
                        NAME

Sometimes it is convenient to indicate the shape and dimensions of
an "isolated section." Section o-o is an isolated section taken on line
drawing of the Try-square and is drawn to twice the size. It is customa
outline of the actual surface cut, and the view may be placed at any cor
sheet of paper as long as it does not interfere with the other views.

1. What is an isolated section?
2. Why is it used?

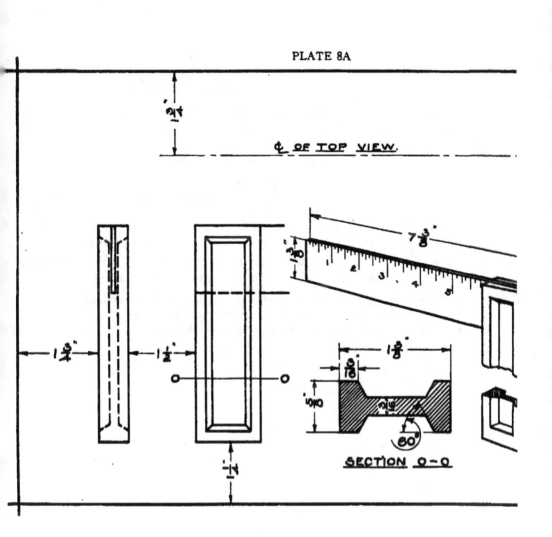

Ȼ OF TOP VIEW.

SECTION O-O

Specification—Plate 9.

Make a complete working drawing of the Nail Box, showing the f and right-side view. All material is ⅜" thick. Scale 12" = 1'.

Title:—   NAIL BOX
            SCALE   DATE
                    NAME

The side of an object represented by one of the views often consists o faces or divided parts. Each of the divisions or details should be dimensi these "detail" dimensions should be equal to the total, or "over-all" dime the object. Both the detail dimensions and the over-all dimension should

1. What is meant by "detail" dimensions? By "over-all" dimensions?
2. Should the detail and over-all dimensions both be indicated?
3. How many board feet of lumber are there in the nail box?

PLATE 9

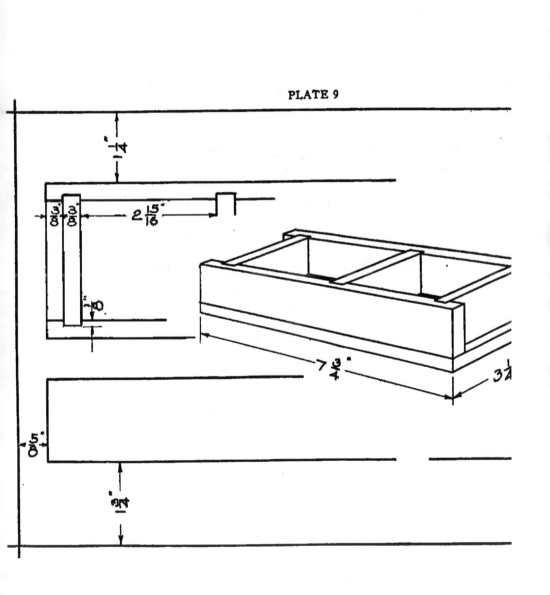

Specification—Plate 9A.

Make a complete working drawing of the Bench-hook, showing the f and right-side view.   Scale 12"=1'.

Title:—   BENCH-HOOK

SCALE   DATE

NAME

1. How many board feet of lumber are required to make 48 bench-hooks?

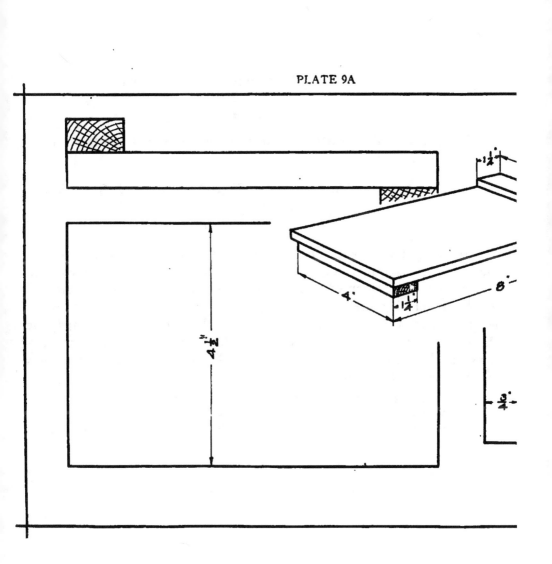

Specification—Plate 10.

Make an assembly drawing of the Mortise-and-Tenon Joint, showing 1 right-side view. Scale 12″—1′.

Title :— MORTISE-AND-TENON JOINT

SCALE   DATE

NAME

Working drawings may be divided into two general classes: assembly drawings. An assembly drawing is a drawing of an object which consists separate parts, represented as they appear when put together, or assemble ing is one which represents each single or separate part of the object. ' cut the stones to proper shape and size, work from detail drawings; the the stones in their proper position works from an assembly drawing. Like who machines the fly-wheel works from a detail drawing; the mechanic w together works from an assembly drawing.

1. What is meant by an "assembly drawing" of an object?
2. What is meant by a "detail drawing" of an object?
3. Why is the right-side view preferred to the left-side view in this case?

PLATE 10

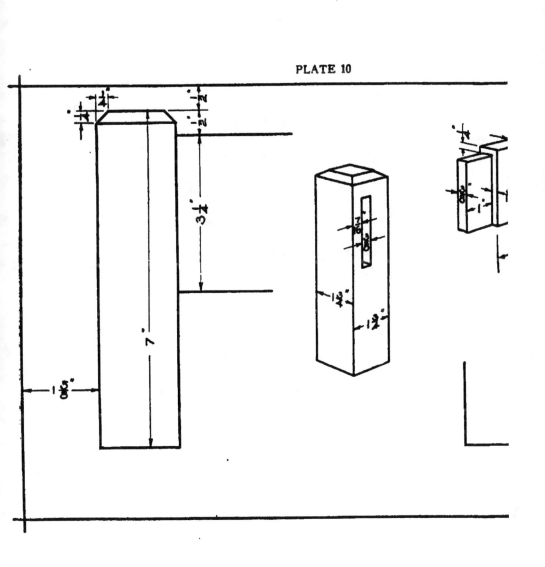

Specification—Plate 10A.

Make a complete working drawing of the Dovetail Joint, showing the fr⟨
⟨iew and right-side view.   Scale 12″=1′.

Title:—   DOVETAIL JOINT
                SCALE   DATE
                        NAME

1. Where have you seen a joint of this kind used?

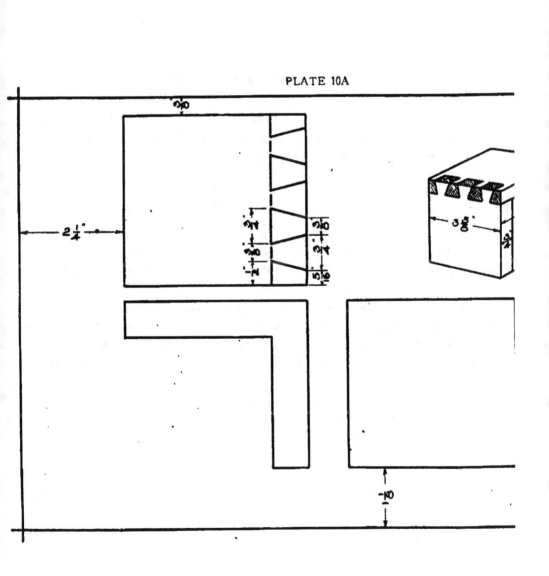

Specification—Plate 11.

Make a complete working drawing of the Ring, showing the front view. The upper half of the side view is to be shown in section. Scale 1⁒

Title :—   RING
          SCALE   DATE
                     NAME

When a view of an object which is symmetrical about an axis (alike center line) is to be shown in section, it is a good plan to draw only one-l view of the object is then said to be "half in section." See page 197.

In making a drawing of a cylindrical or circular object all center lin first. The view showing the circular form of the object should then be drawi or views projected from it. Centers for arcs should always be located wit and all circles and arcs should be drawn first in inking. The character ₵ for the term "center line."

1. When is it a good plan to show a view of an object "half in section"?
2. Outline the steps in drawing the Ring.
3. Why should arcs and circles be inked first?

PLATE 11

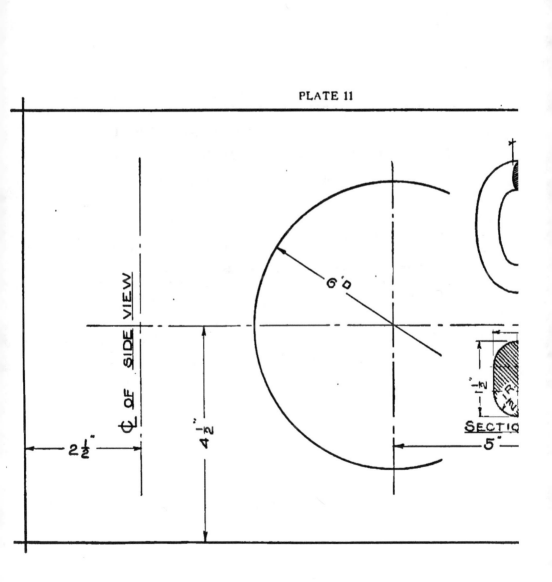

C OF SIDE VIEW

$2\frac{1}{2}$"

$4\frac{1}{2}$"

6"∅

$1\frac{1}{2}$"

R-$1\frac{1}{2}$"

SECTIO

5"

Specification—Plate 11A.

Make a complete working drawing of the Strap.   Scale 12″=1′.

Title :—   STRAP

SCALE   DATE

NAME

NOTES ON DIMENSIONING.

1. A center line should never be used as a dimension line.
2. The radius of an arc of a circle should be marked R, or RAD.
3. In locating holes, always indicate the distance between center lines.

PLATE 11A

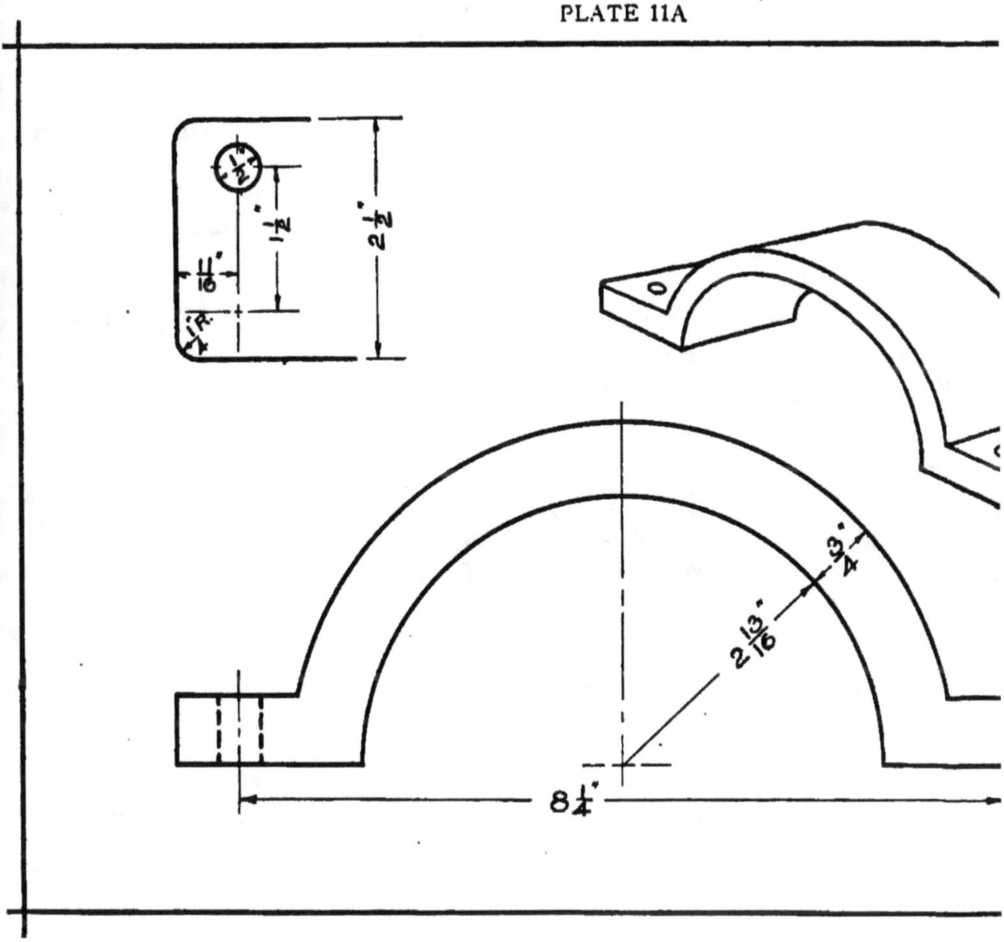

Specification—Plate 12.

Make a complete working drawing of the Face-plate. Bore ⅞". Scale

Title:—   FACE-PLATE

SCALE   ·   DATE

NAME

Sharp inside corners make a casting or a machine part weak, and should
if possible. These corners may be strengthened by making them rounded.
called a "fillet." The corners are said to be "filleted."

In a drawing, fillets are constructed with circular arcs. The radius (
always be indicated.

1. What is a tangent of a circle? What is meant by the "point of tangency"?
2. What is an arc? A chord? A segment?
3. Why are inside corners in a casting "filleted"?

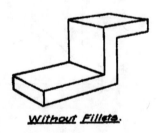

Without Fillets.

Filleted.

PLATE 12

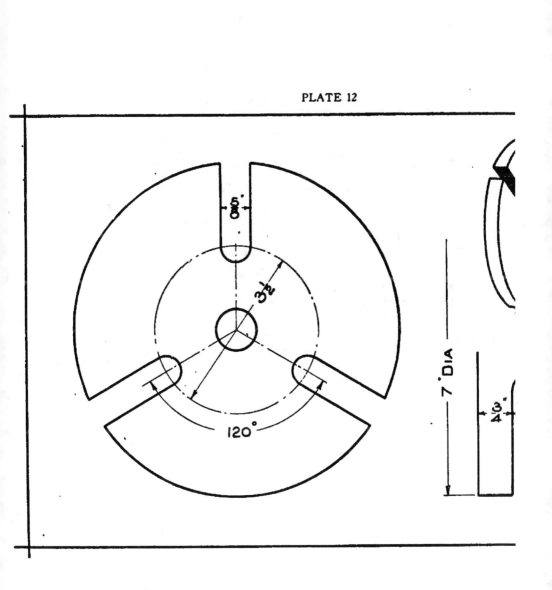

Specification—Plate 12A.

Make a complete working drawing of the Valve Stem. Scale 12″=1′

Title:—   VALVE STEM

SCALE   · DATE

NAME   .

NOTES ON DIMENSIONS.

1. Always give the diameter of a circle, not the radius.
2. Never omit the size of fillets.
3. Do not forget center lines.

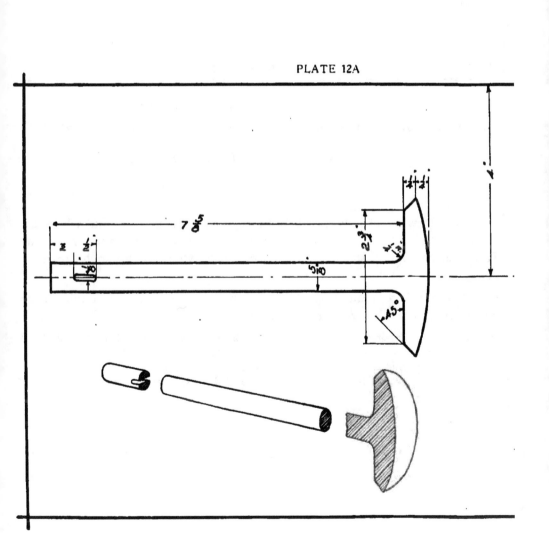

Specification—Plate 13.

Make a complete working drawing of the Clutch Thimble.   Scale 12"-

Title :—   CLUTCH THIMBLE

SCALE   DATE

NAME

In a drawing of a cylindrical object of many diameters and where on sists of many circles, it is not good practice to place the diameter dimens It is better to put them in the other view where they can be read an easily.

A dimension should never be placed on a center line and should never space which is too narrow for the figures.

1. When the drawing of an object involves many circles, where should the d placed?   Why?
2. Why should dimensions not be placed on center lines?
3. Why should a dimension never be crowded into a narrow space?

PLATE 13

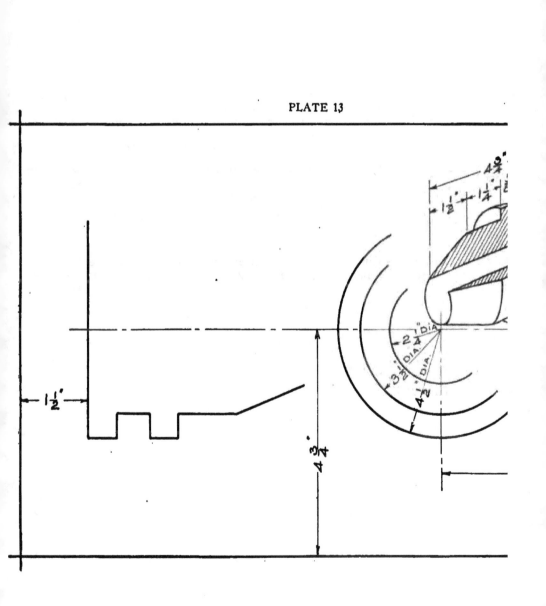

Specification—Plate 13A.

Make a complete working drawing of the Shaft Bracket.   Scale 12″—1ʹ

Title:—   SHAFT BRACKET

SCALE   DATE

NAME

1.   Why is it unnecessary to draw three views of the Shaft Bracket?
2.   Why is a drawing with the front and top view preferred to a drawing with th

PLATE 13A

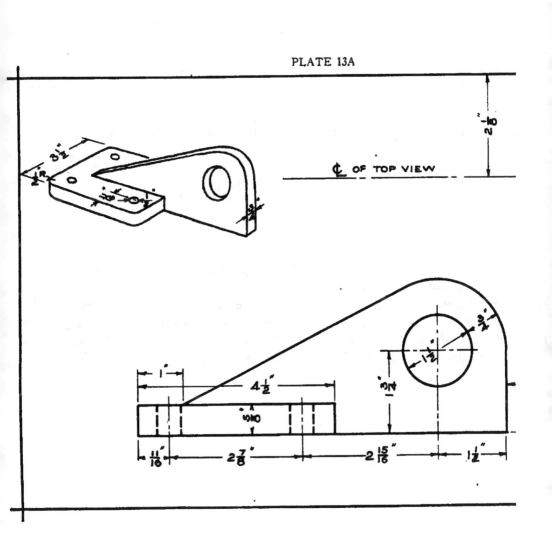

$\mathbb{C}$ OF TOP VIEW

Specification—Plate 14.

Make a complete working drawing of the Pen Tray, showing the front the right-side view in section.   Scale 9″=1′.

Title :—   . PEN TRAY

SCALE   DATE

NAME

KEEP THE RULING PEN CLEAN.

Drawing ink consists of a black pigment held in solution with an acid. rates quickly so that lines drawn with the ink will dry rapidly.   The ruling be cleaned constantly as the ink becomes thick and sluggish between the When this occurs the ink does not flow freely and the result is a ragged lin While inking, a soft cloth should be used occasionally to wipe out the thic the nibs clean and bright both inside and outside.

1. Why should a ruling pen be kept clean?

PLATE 14

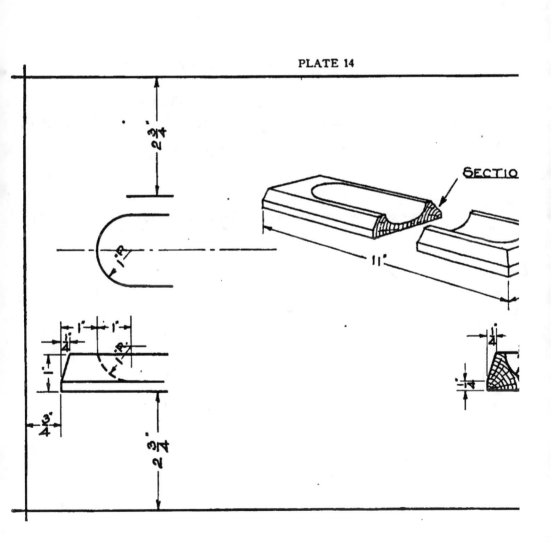

Specification—Plate 14A.

Make a complete working drawing of the Pin Bearing. Scale 12″=

Title :—   PIN BEARING

         SCALE   DATE

              NAME

1. Why are the front and side views preferred to the front and top views in t
Bearing?

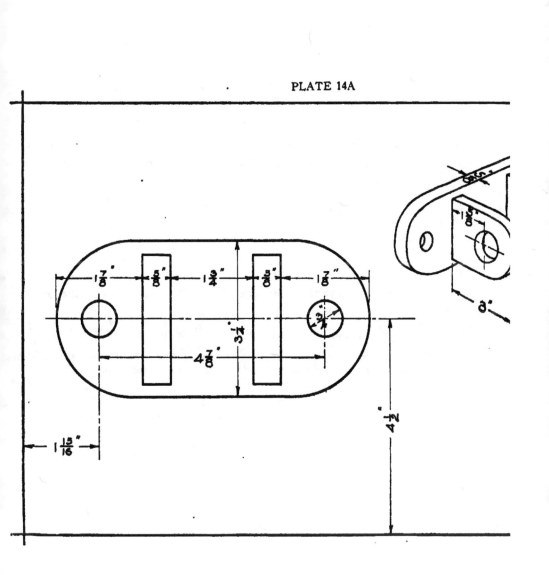

Specification—Plate 15.

Make an assembly drawing of the Book-rack showing two views, should be shown and all parts dimensioned. In the picture drawing one show the construction. Scale 6″=1′.

Title:—   **BOOK-RACK**

SCALE   DATE

NAME

A "bill of material" for an object to be constructed of wood, is a t which gives the number wanted, the size, the material and the name of eac make it.

Lumber sizes should be given in the following order: Thickness by

1. Make out a bill of material for the Book-rack using the form as shown below.
2. How many board feet of lumber are required to make the Book-rack?

BILL OF MATERIAL FOR PLANT STAND.

4 pcs. 1¼″ x 1¼″ x 22″   Birch for Legs
4 "   ¾″ x 2″   x 10½″   "   " Rails
1 "   ¾″ x 14″   x 14″   "   " Top

PLATE 15

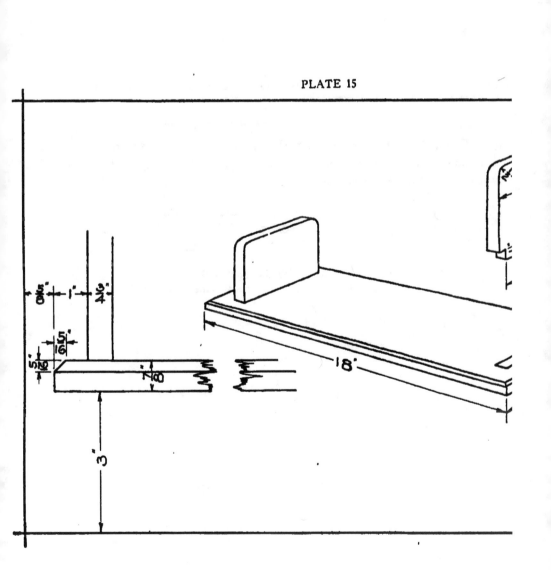

Specification—Plate 15A.

From the detail drawings make an assembly drawing of the Footstool and side views. The views are to be located so that they will appear well ba within the border lines. Scale 6″—1′.

Title:—   FOOTSTOOL

    SCALE   DATE

        NAME

1. Make out a bill of material for the Footstool.
2. How many board feet are required to make the Footstool?

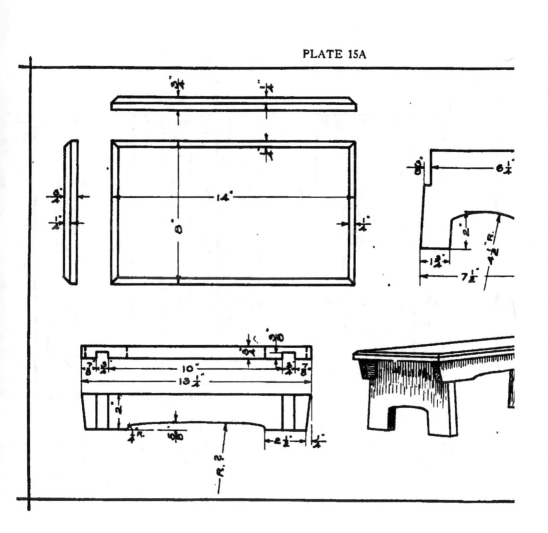

TABLE OF STROKES

	STROKES		
	1	2	3
l	l		
i	l	i	
t	l	t	
z	‑	7	z
v	|	v	
w	|	v	w
x	\	x	
y	|	y	
k	l	k	k

	STROKES		
	1	2	3
j	J	j	
f	l	r	f
r	l	r	
h	l	h	
n	l	n	
m	l	n	m
u	c	u	
o	c	o	
c	c		

	ST
	1
e	c
a	c
d	c
g	c
b	l
p	l
s	s
8	s
2	2

Slant of letters about 70°

PLATE 16

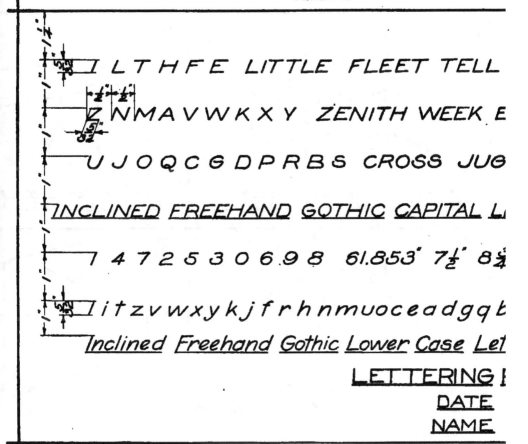

ILT HFE LITTLE FLEET TELL

ZNMAVWKXY ZENITH WEEK E

UJOQCGDPRBS CROSS JUG

INCLINED FREEHAND GOTHIC CAPITAL L

1 4 7 2 5 3 0 6.9 8 61.853° 7½″ 8¾

litzvwxykjfrhnmuoceadgqb

Inclined Freehand Gothic Lower Case Let

LETTERING

DATE

NAME

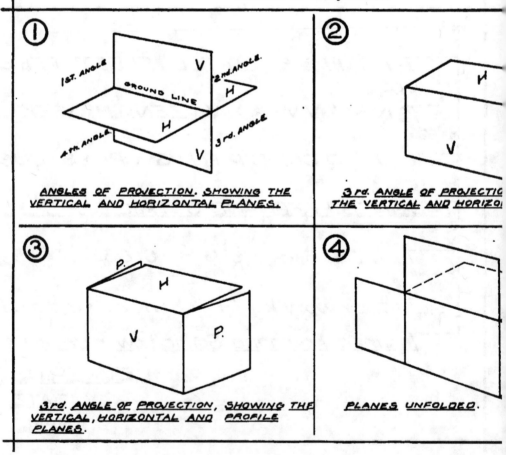

① ANGLES OF PROJECTION. SHOWING THE VERTICAL AND HORIZONTAL PLANES.

② 3rd. ANGLE OF PROJECTION, THE VERTICAL AND HORIZON...

③ 3rd. ANGLE OF PROJECTION, SHOWING THE VERTICAL, HORIZONTAL AND PROFILE PLANES.

④ PLANES UNFOLDED.

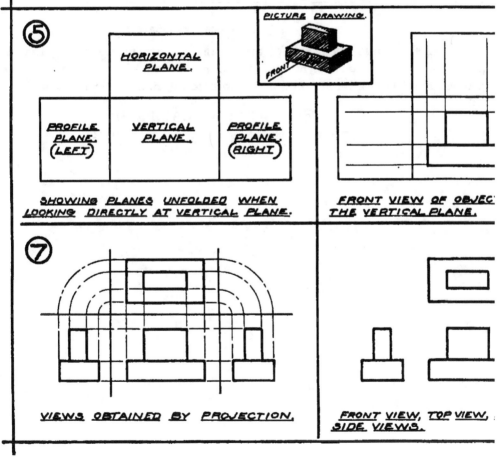

⑤

HORIZONTAL PLANE.

PICTURE DRAWING.

FRONT

PROFILE PLANE. (LEFT)

VERTICAL PLANE.

PROFILE PLANE. (RIGHT)

SHOWING PLANES UNFOLDED WHEN LOOKING DIRECTLY AT VERTICAL PLANE.

FRONT VIEW OF OBJEC THE VERTICAL PLANE.

⑦

VIEWS OBTAINED BY PROJECTION.

FRONT VIEW, TOP VIEW, SIDE VIEWS.

Specification—Plate 17.

In this problem two views of the Triangular Prism are given; the vie
plane and the view on the horizontal plane. Obtain the view on the rigl
projection. Show all construction and projection lines.   Scale 12″=1′.

Title :—   TRIANGULAR PRISM

SCALE   DATE

NAME

PLATE 17

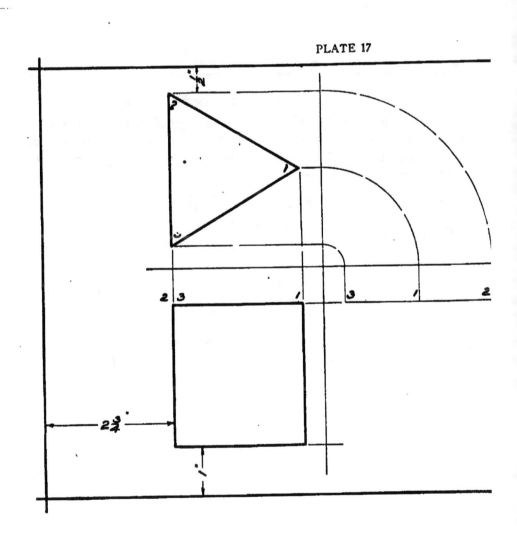

Specification—Plate 18.

In this problem the view of the Hexagonal Prism is given on the hori
view on the vertical plane is partially drawn and should be completed by prc
on the right profile plane should be obtained by projection. Show all con
jection lines. Scale 12"—1'.

Title:— HEXAGONAL PRISM

SCALE   DATE

NAME

PLATE 18

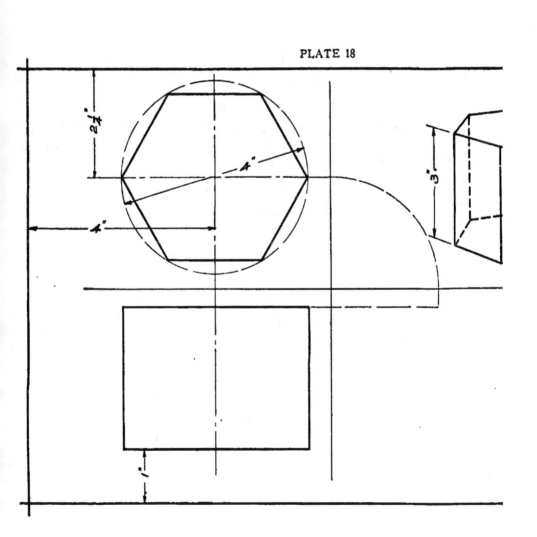

Specification—Plate 19.

The view on the vertical plane and the view on the horizontal plane are
the view on the right profile plane by projection. Show all constructic
lines. Scale 12″—1′.

Title:— OCTAGONAL PRISM

SCALE   DATE

NAME

PLATE 19

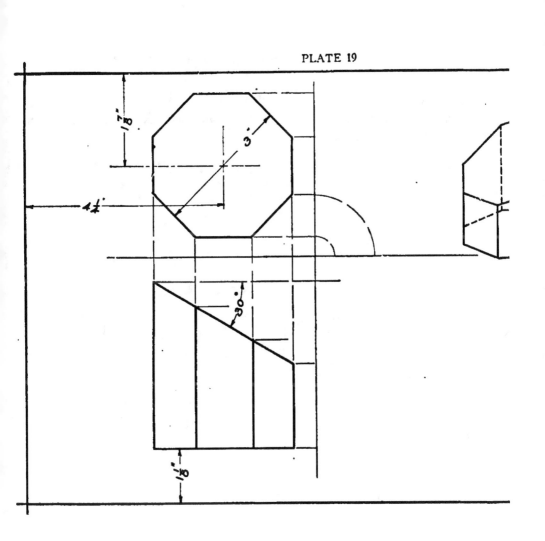

Specification—Plate 20.

Obtain the view on the right profile plane by projection.　Show all projection lines.　Scale $12''=1'$.

Title:— TRIANGULAR PYRAMID

SCALE　　　　　　　　　　　　DATE

NAME

PLATE 20

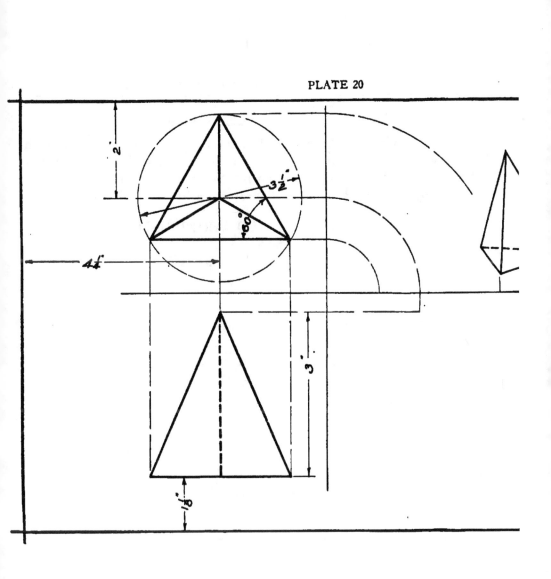

Specification—Plate 21.

Draw two views of the Square Prism on the left of the sheet, as sh
side of the sheet draw three views of the prism when tilted as indicated
vertical plane and the view on the horizontal plane are complete. Obtain
file plane by projection. Show all projection lines. Scale 12″—1′.

Title:—   SQUARE PRISM

SCALE   DATE

NAME

PLATE 21

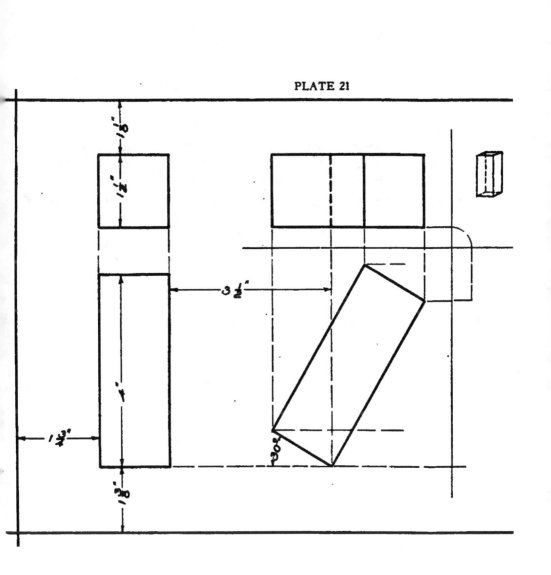

Specification—Plate 22.

Draw two views of the Triangular Prism on the left side of the s
the right side of the sheet draw three views of the prism when tilted as
on the vertical plane and the view on the left profile plane are complete.
the horizontal plane by projection. Show all projection lines. Scale 12′

Title :—   TRIANGULAR PRISM

SCALE   DATE

NAME

PLATE 22

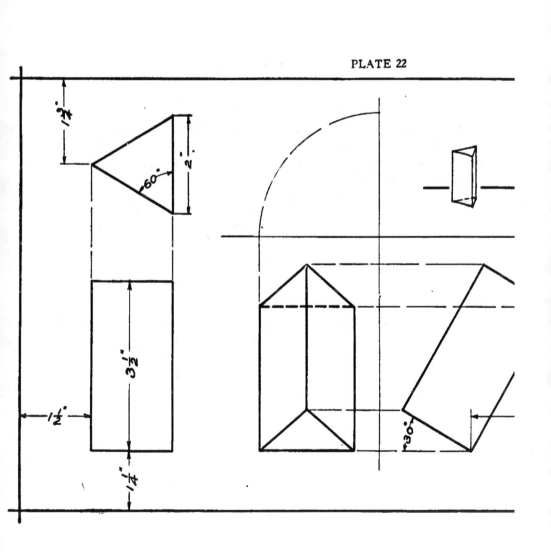

Specification—Plate 23.

Draw two views of the Hexagonal Pyramid on the left side of ꜰ On the right side of the sheet draw three views of the pyramid when tilt view on the vertical plane is complete. Obtain the view on the horizont on the right profile plane by projection. Scale 12″—1′.

Title:— HEXAGONAL PYRAMID
               SCALE   DATE
                       NAME

PLATE 23

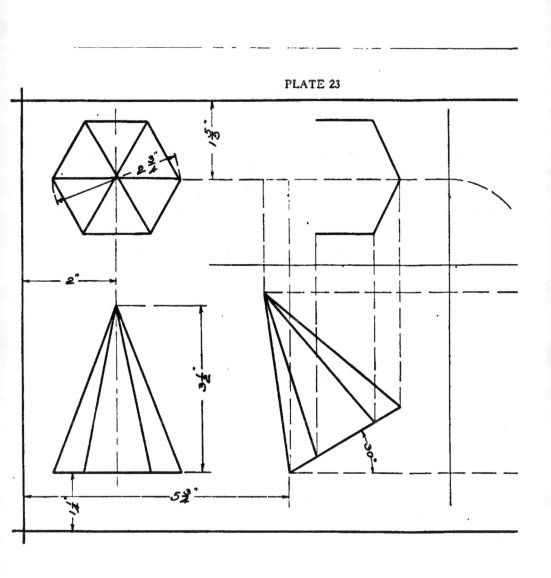

Specification—Plate 24.

Draw two views of the H-block on the left side of the sheet, as show of the sheet draw two views of the block as it would appear when turne(view on the horizontal plane is partially drawn and makes an angle of plane. Obtain the view on the vertical plane by projection. Show all pro $12'' = 1'$.

Title:—   H-BLOCK

SCALE   DATE

NAME

PLATE 24

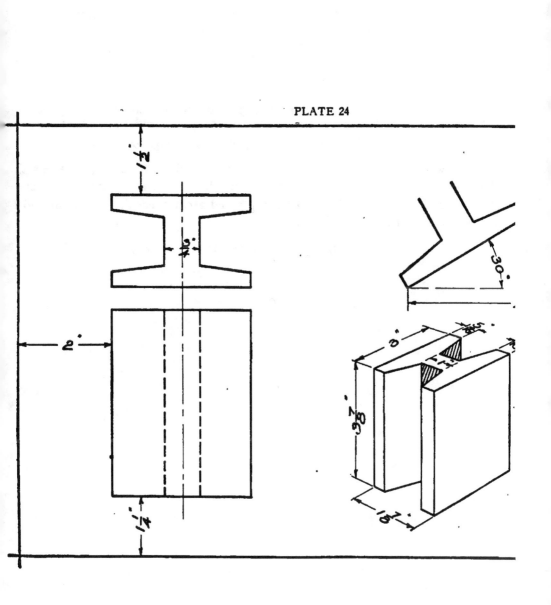

Specification—Plate 25.

Draw two views of the Notched Block on the left side of sheet, as sh
dle of sheet draw two views of the block when tilted as indicated. On the
draw two views of the block when it is tilted as in preceding problem and t
angle of 30° with the vertical plane.

Title:—   NOTCHED BLOCK

SCALE   DATE

NAME

PLATE 25

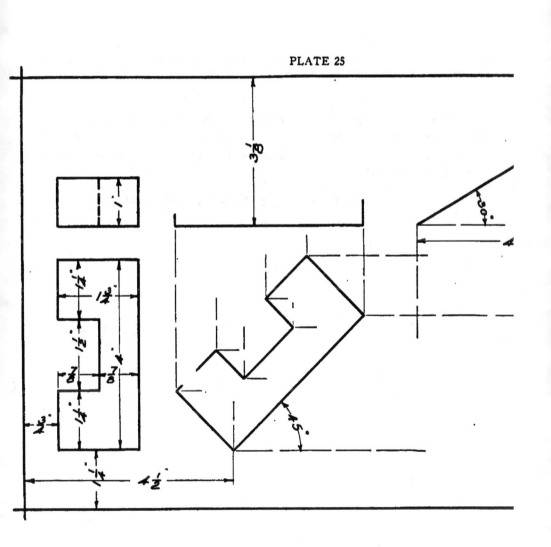

Specification—Plate 26.

Draw three views of the Truncated Pyramid. The pyramid is so edge of the base is parallel to the vertical plane. It is cut by an imaginar an angle of 45° to the horizontal plane and 90° to the vertical plane. lines. Scale 12"—1'.

Title:— TRUNCATED PYRAMID
SCALE   DATE
NAME

PLATE 26

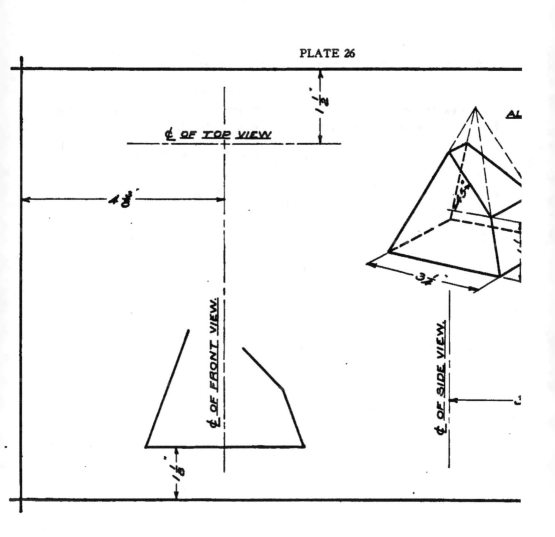

¢ OF TOP VIEW

¢ OF FRONT VIEW.

¢ OF SIDE VIEW.

Specification—Plate 26A.

Draw three views of the Truncated Prism when placed as indicated shown. Show all projection lines. Scale 12″=1′.

Title:—   TRUNCATED PRISM

SCALE   DATE

NAME

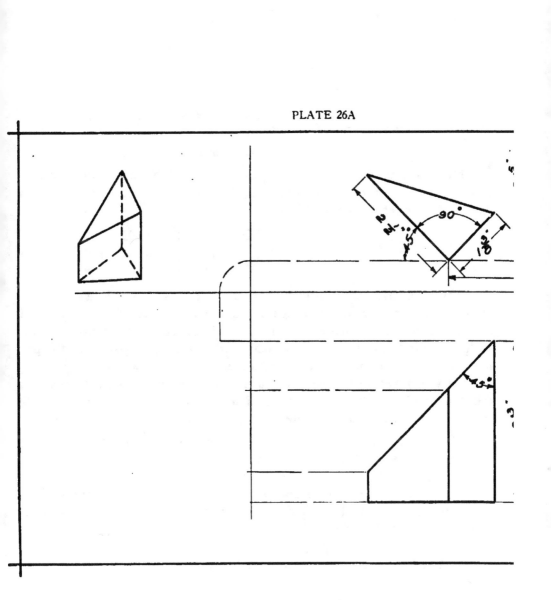

Specification—Plate 27.

Make a complete working drawing of the Tool-post Slide, showing right-side view. Indicate with a note that the piece is to be made of ca "finished all over." Scale 12″=1′.

Title :—   TOOL-POST SLIDE
          SCALE   DATE
              NAME

The production of an ordinary metal casting, as iron or brass, involves ations. First, a form or model very nearly like that required in the casti or some other easily shaped material, and is called the "pattern." Next, mold is made in sand. The sand, which is especially prepared for the pur ing sand. It contains a small percentage of clay and will retain its shap has been removed. It is held in place while the mold is being made b called a "flask." The flask also facilitates making the proper cavity in th tern. Lastly, the metal is melted and poured into the mold.

The surfaces of a casting which are made true and smooth by cutting are called "finished" surfaces. Such surfaces must have added metal a the casting so that the casting will be the required size after the surfaces h A drawing should always be made and dimensioned to the required m: object to be made. All finished surfaces must be labelled so that due made in the pattern. See page 195.

1. How are metal castings made?
2. What is meant by "finish" in a working drawing?
3. What is meant by "finish all over"?

PLATE 27

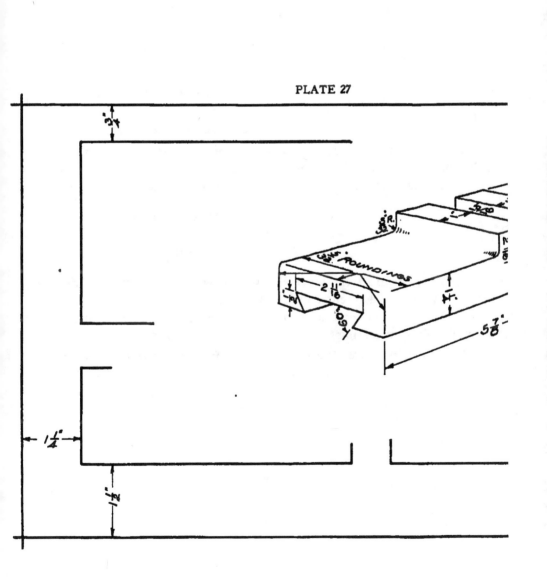

Specification—Plate 27A.

Make a complete working drawing of the Hexagonal Wrench shov
a "revolved section" of the handle. The two faces of the wrench are to b
by a note that the wrench is to be made of malleable iron. Scale 12″—1

Title:— HEXAGONAL WRENCH
SCALE   DATE
NAME

Ordinary iron castings are brittle and will break rather than bend
severe strain. The wrench in this problem is cast of iron and then put th
which makes it tougher than ordinary iron castings. The product is "ma
a "malleable iron casting."

1. How does a malleable iron casting differ from an ordinary iron casting?

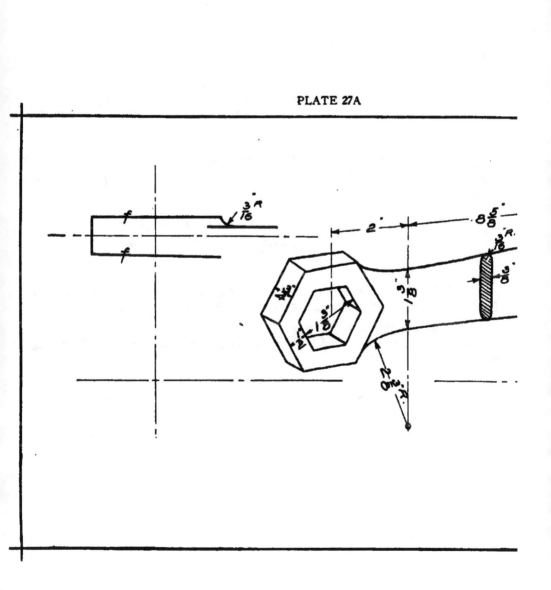

Specification—Plate 28.

Make a drawing of the Picture Frame, showing the front view and th
tion.   The full-size detail shows a section of the molding which is used t
Scale 6″=1′.

Title :—   PICTURE FRAME

SCALE   DATE

NAME

In small picture frames the joints are generally glued and nailed.   Hol
for the nails while the pieces are held together with clamps.

1.   What is meant by a "mitre-joint"?

2.   How many feet of molding are necessary to make the frame in the drawing,
stock?

PLATE 28

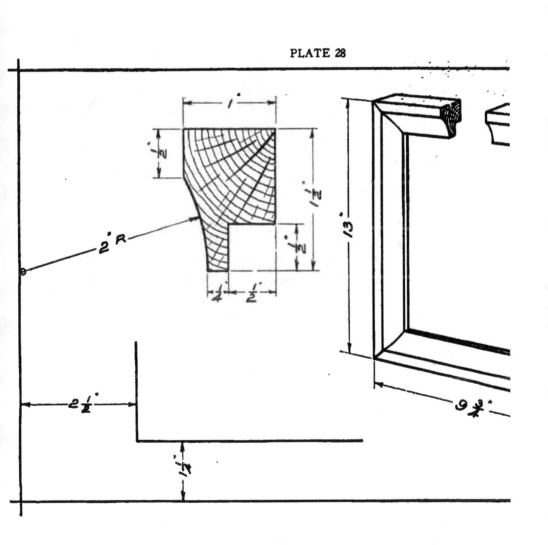

Specification—Plate 28A.

Make a drawing of the Nut Bowl, showing the front-view half in sect
half of the top view. Scale 12″=1′.

Title:—   NUT BOWL

SCALE   DATE
NAME

When a cylindrical object is symmetrical, or the same on both sides of
sometimes a saving of time and paper to draw only one-half of the profil
This is especially true when the object involves many small details.

1. Why is one-half of the top view omitted in the drawing of the Nut Bowl?

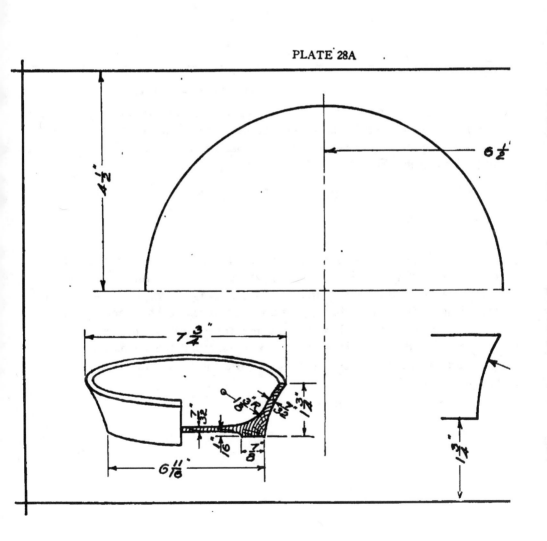

Specification—Plate 29.

Make a working drawing of the Gland consisting of two views. Show tl section. The drawing in the lower left-hand part of the data sheet shows a fu bevels on the gland. The total length of gland is 2¾″. Indicate by a note tha finished all over. Scale 12″=1′.

Title :—   GLAND

SCALE   DATE

NAME

The gland is used to retain and compress the packing in a stuffing b tion.) A stuffing box in a steam engine is a piece of mechanism at the of the piston-rod enters the cylinder. By screwing down the nuts on the st forced deeper into the stuffing box and compresses the packing against tl the walls of the stuffing box, thereby preventing leakage of steam from the cylinder thru the stuffing box and around the piston-rod. The packing being a fibrous material, allows the rod to move back and forth freely.

1. Explain the use of the gland.

PLATE 29

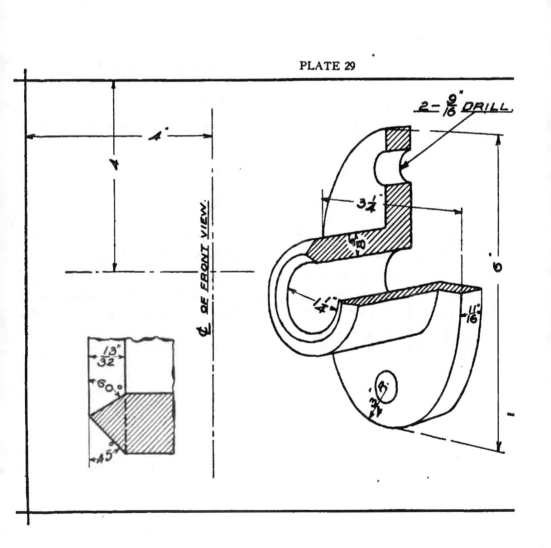

Specification—Plate 29A.

Make a drawing of the Gear Blank, showing the side view in full s
= 1″.

Title:—   GEAR BLANK

SCALE   DATE

NAME

A gear wheel is a circular disk or wheel with cogs or gear teeth cut i
used to transmit motion by means of engaging with other gear wheels or

The gear blank in the drawing has a rectangular groove, or "keyway"
A similar keyway is cut into the shaft on which the gear is mounted. A "l
into the keyways, locks the gear on the shaft so that gear and shaft revolv

1. Where have you seen a gear wheel in use?
2. How can a wheel which is mounted on a shaft be fastened to the shaft so th

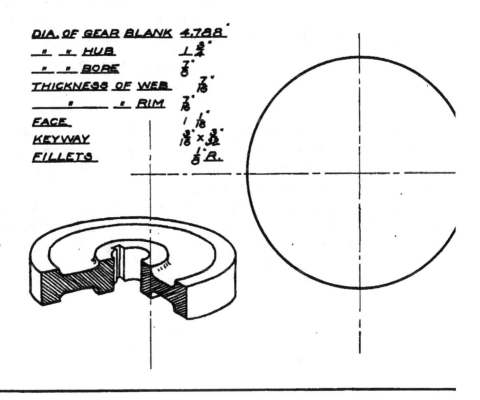

DIA. OF GEAR BLANK 4.788"
"   " HUB $1\frac{3}{4}$"
"   " BORE $\frac{7}{8}$"
THICKNESS OF WEB $\frac{7}{16}$"
"   " RIM $\frac{7}{8}$"
FACE $1\frac{1}{8}$"
KEYWAY $\frac{3}{16}$" × $\frac{3}{32}$"
FILLETS $\frac{1}{8}$" R.

Specification—Plate 30.

Make a working drawing of the 5″ Pipe Elbow, showing the front
Show a partial section in the side view as suggested in the picture draw
sary to show invisible lines representing the drilled holes. Scale 12″=1′.

Title:—   5″ PIPE ELBOW

SCALE   DATE

NAME

The 90° pipe elbow is one of many so-called "pipe fittings" which
connect pipes. The smaller sizes of pipe usually have screwed joints. T
on the ends and connected by threaded fittings. The larger sizes of pi
nected by flanged fittings bolted together.

The inside diameter of a pipe is the nominal diameter. In speaking
is understood that the inside diameter is referred to; the inside diameter
compute the amount of gas that may pass through the pipe.

1. How are pipes connected?
2. Why is the inside diameter of a pipe given as the nominal diameter?

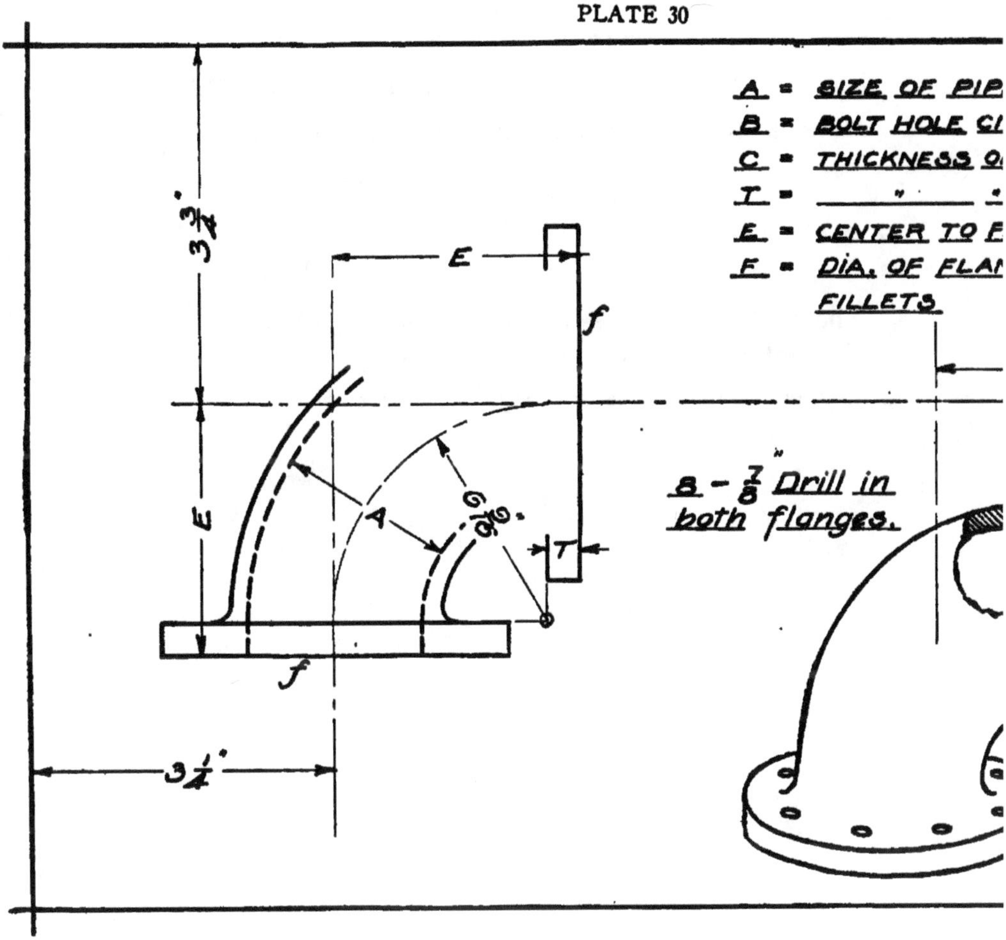

PLATE 30

A = SIZE OF PIP

B = BOLT HOLE CI

C = THICKNESS O

T =   "   :

E = CENTER TO F

F = DIA. OF FLAI

FILLETS

8 - 7/8" Drill in both flanges.

E

A

90°

T

f

f

3 3/4"

3 1/4"

Specification Plate 30A.

Make a drawing of the Chain as indicated. Scale 12″=1′.

Title:—   CHAIN

SCALE   DATE

NAME

The area of a circle = radius x radius x 3.1416.
The volume of a cylinder = area x length.

1. Compute the volume of a single link of the chain.

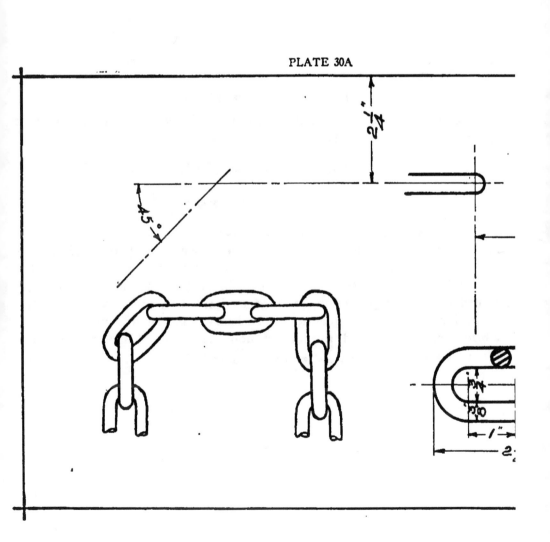

Specification Plate 31.

Make a drawing of the Babbitt Bearing, indicating the materials by mea
one of the views.   Scale 12″=1′.

Title:—   BABBITT BEARING
SCALE   DATE
NAME

The bearing in the drawing is used to support a revolving shaft.   As i
machine parts where motion and power are transmitted, there is friction be
of the moving part and the surface of the part that bears or supports it.   Thi
loss of power, and causes wear on the surfaces of the moving part and the su
In order to minimize friction, bearings are lined with one of several kinds c
which have anti-friction qualities.   The bearing in the drawing is lined w
which is an alloy of lead, tin and antimony and is one of the most common

1.   What is babbitt metal and what is it used for?

PLATE 31

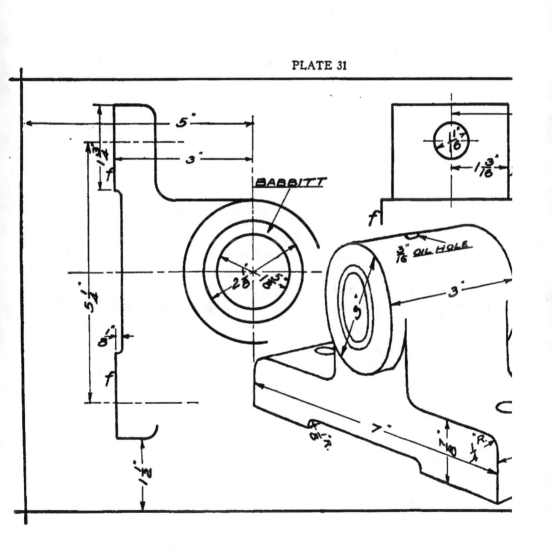

BABBITT

OIL HOLE

Specification—Plate 31A.

Complete the drawing of the Flanged Bushing. The bushing is finisl
$12'' = 1'$.

Title:—   FLANGED BUSHING
SCALE   DATE
NAME

A bushing is a lining, or tube of metal or other material which is insei
has been drilled or bored. It is used to reduce the size of the hole or to
bearing surface. The bushing in the drawing is made to fit the end be
case of a gas engine and forms a bearing surface for the revolving crank
bearing becomes worn thru continued use, it can be replaced by a new o
crank shaft.

1. What is a bushing?
2. Where have you seen a bushing used?

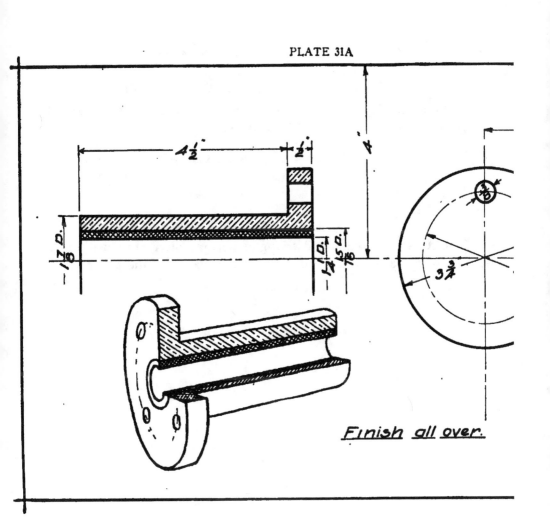

Finish all over.

Specification—Plate 32.

Make a drawing of the Rocker Arm as indicated. Place fillets where the radius of all fillets ⅛″. Scale 9″—1′

Title:— ROCKER ARM

SCALE DATE

NAME

KEEP YOUR TOOLS CLEAN.

The tee-square, scale and triangles become dirty thru use and shoul sionally with warm water and soap. Dirty tools soil the paper and shoul

PLATE 32

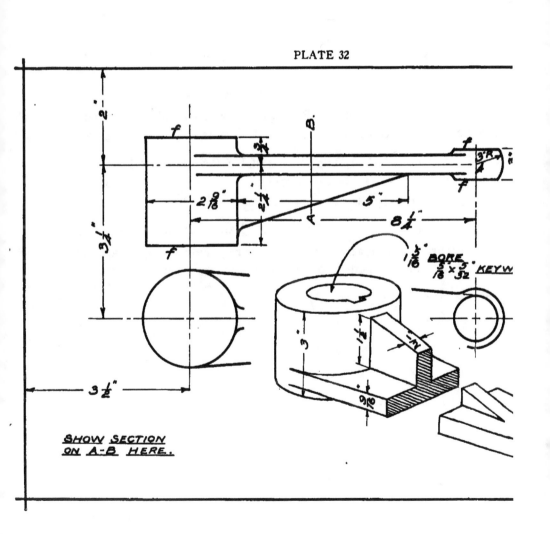

SHOW SECTION
ON A-B HERE.

Specification—Plate 32A.

Make a drawing of the Bell-crank Lever as indicated. All fillets ¼
specified. Diameter of large hub 3¼". Diameter of small hubs 2¼". Scal

Title:—   BELL-CRANK LEVER

SCALE   DATE

NAME

Keep Your Pencil Points Sharpened.

Accurate and neatly executed drawings require well sharpened lead po
lead should be sharpened to a chisel point similar to the chisel point on a

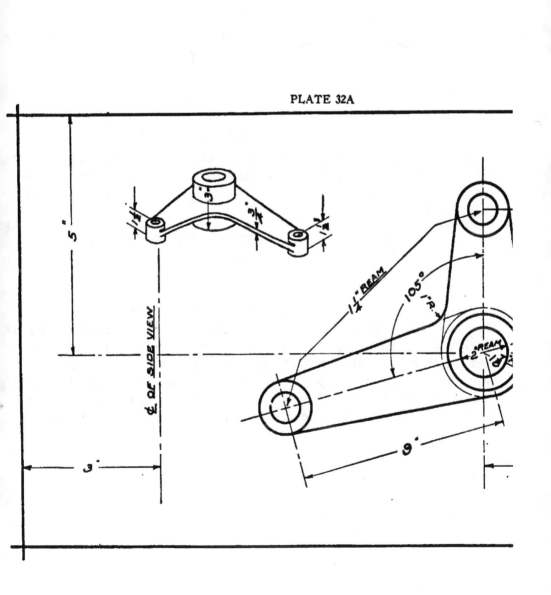

Specification—Plate 33.

The working drawing represents a Truncated Square Prism, showing t
view, side view and an auxiliary, or added view. The auxiliary view sho\
size of the surface which is cut by an oblique plane. Develop a pattern for
cated. Scale 12″═1′.

Title:—   TRUNCATED SQUARE PRISM
           SCALE   DATE
              NAME

THE DEVELOPMENT OF SURFACES.

It is frequently necessary to make a drawing of the surfaces of an o\
such a manner that a pattern being made from it and properly folded or ro\
duce the object. In order to do this, an outline of each surface must be ob\
of projection parallel to it, so that it will be represented in its true shape a\
faces must then be grouped and drawn adjacent to each other so that th\
would assume the exact shape and size of the object, if properly folded or

PLATE 33

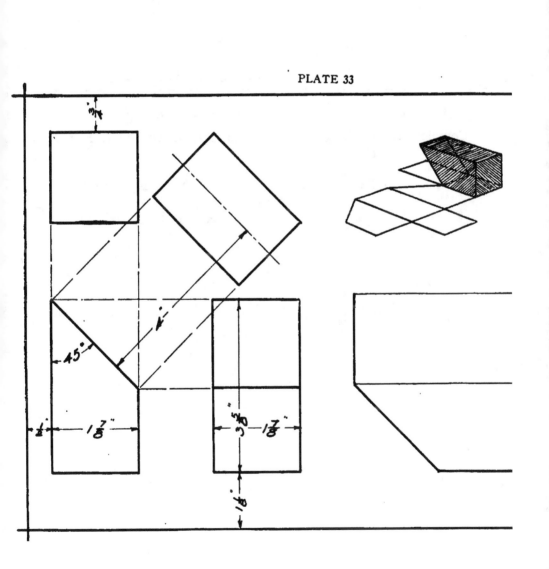

Specification—Plate 33A.

Complete the working drawing of the Truncated Square Prism and same. Scale 12″=1′.

Title:—   TRUNCATED SQUARE PRISM
                 SCALE   DATE
                          NAME

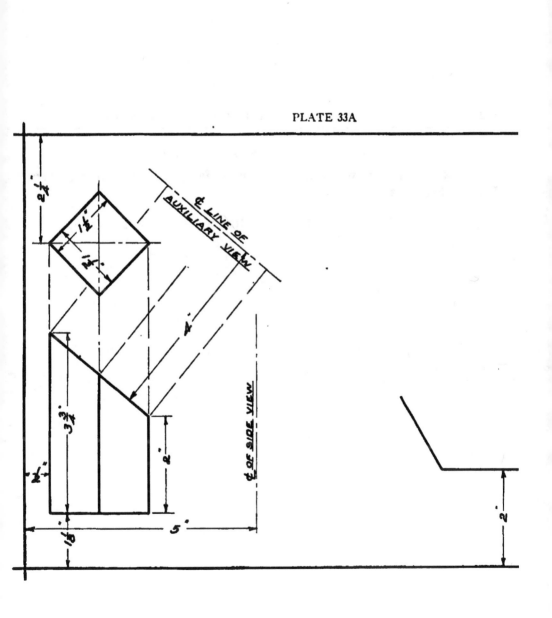

Specification—Plate 34.

Make a drawing, including an auxiliary view, and develop a pattern Prism. Scale 12″—1′.

Title:—   TRUNCATED OCTAGONAL PRISM

SCALE   DATE

NAME

PLATE 34

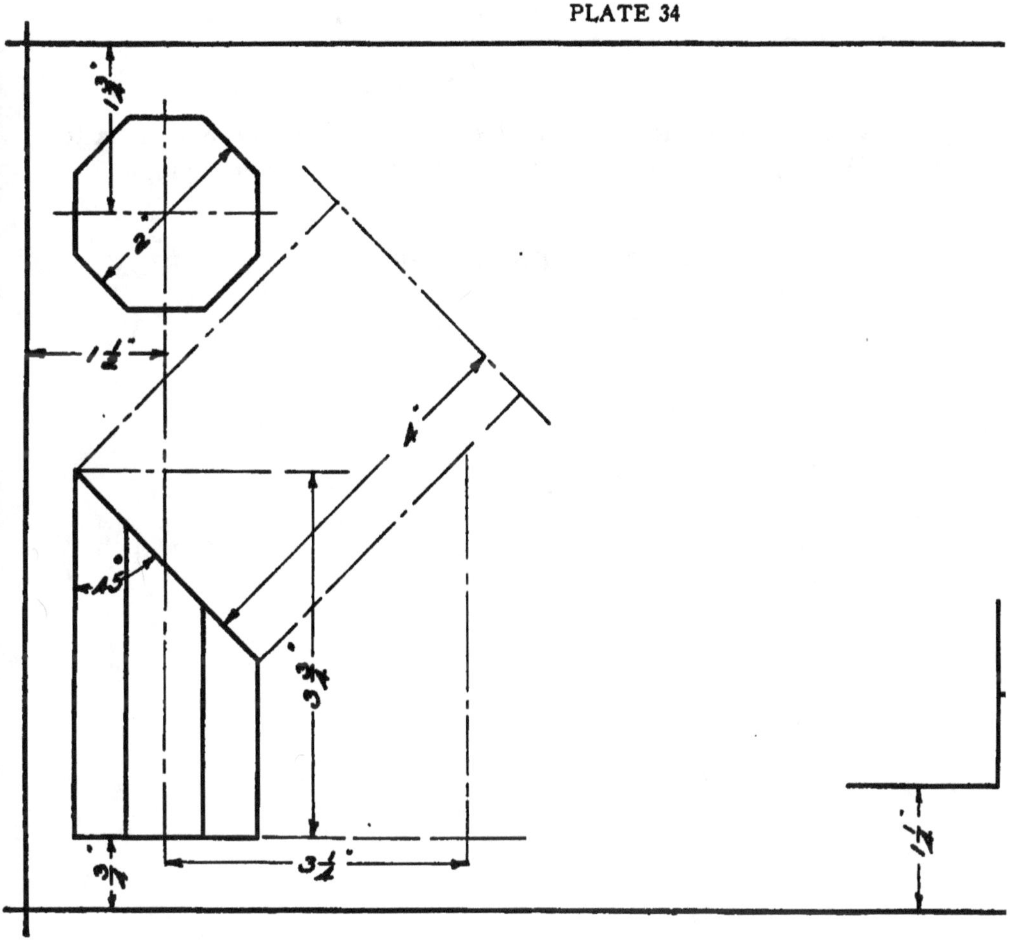

Specification—Plate 34A.

Make a drawing, including an auxiliary view, and develop a pattern Prism. A simple method of drawing a pentagon, having giving one side, 12″—1′.

Title:—   TRUNCATED PENTAGONAL PRISM

SCALE   DATE

NAME

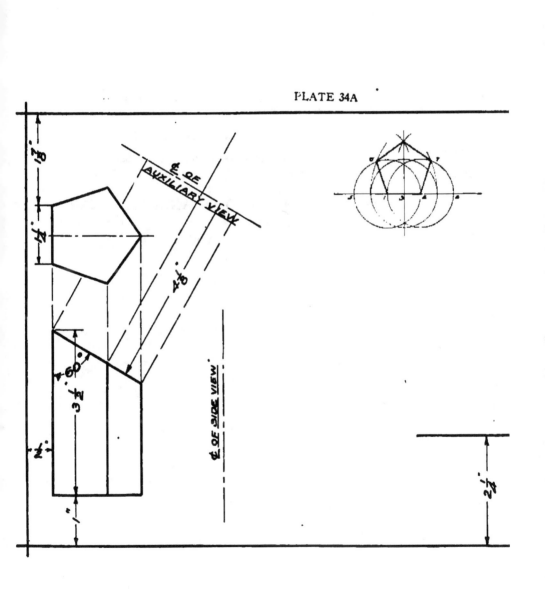

Specification—Plate 35.

Make a drawing, including an auxiliary view, and develop a pattern Prism. The base of the prism is an equilateral triangle. Scale 12″=1′.

Title:—   TRUNCATED TRIANGULAR PRISM

SCALE   DATE

NAME

PLATE 35

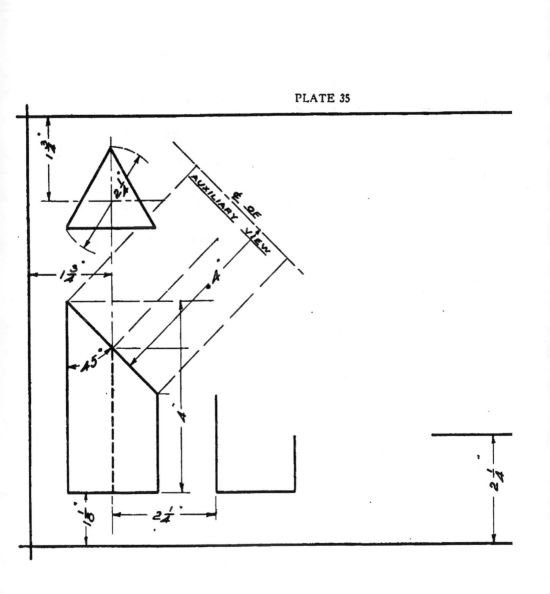

Specification—Plate 35A.

Make a drawing, including an auxiliary view, and develop a pattern Prism. Scale $12'' = 1'$.

Title:—   TRUNCATED TRIANGULAR PRISM

SCALE   DATE

NAME

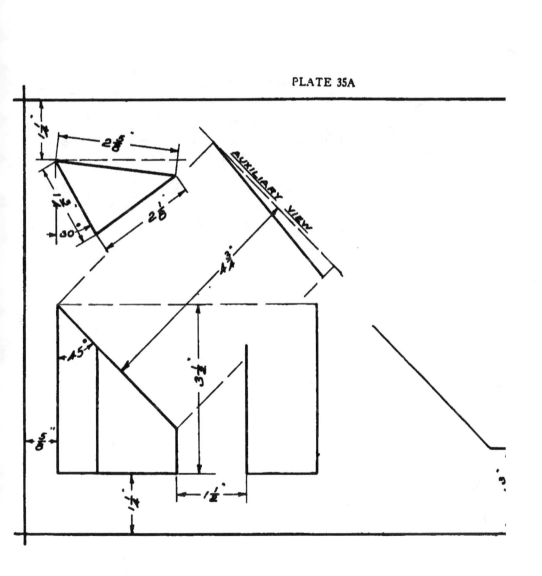

Specification—Plate 36.

Make a drawing, including an auxiliary view, and develop a pattern
Pyramid. The base of the pyramid is an equilateral triangle. Scale 12″—

Title:—   TRUNCATED TRIANGULAR PYRAMID

SCALE   DATE

NAME

PLATE 36

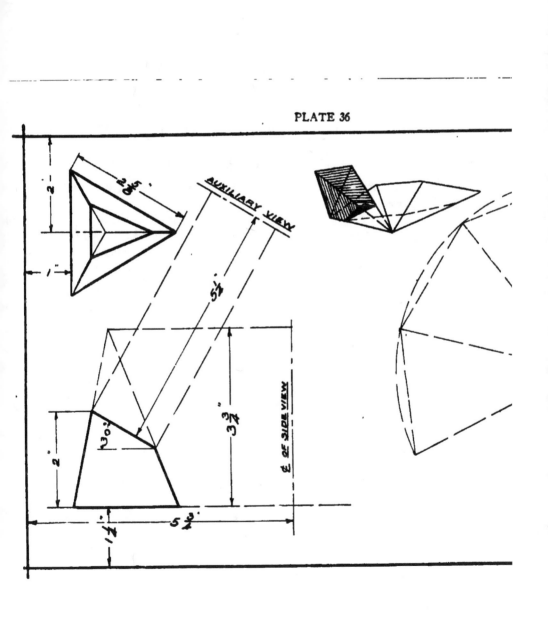

AUXILIARY VIEW

$2\frac{5}{8}$"

2

1"

$5\frac{1}{2}$"

$3\frac{3}{4}$"

30°

2"

$5\frac{3}{4}$"

$1\frac{1}{4}$"

℄ OF SIDE VIEW

Specification—Plate 37.

Make a drawing, including an auxiliary view, and develop a pattern fo mid. Scale 12″=1′.

Title:—   TRUNCATED SQUARE PYRAMID

SCALE   DATE

NAME

PLATE 37

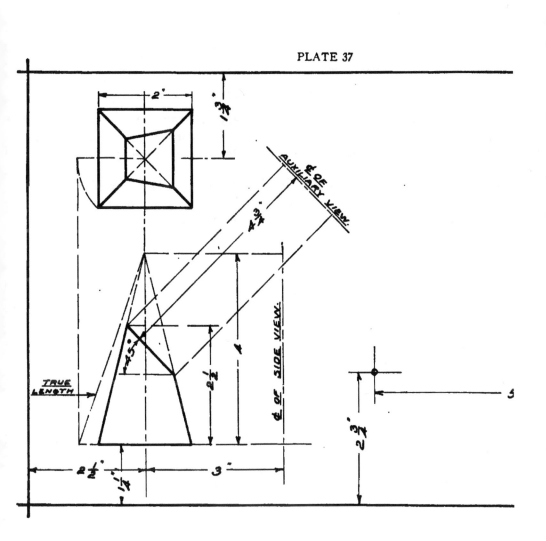

Specification—Plate 37A.

Make a drawing, including an auxiliary view, and develop a pattern for mid. In laying out the pattern draw an arc for each edge of the pyramid equal the true length of the respective edge. Scale 12"—1'.

Title:—   TRUNCATED OBLIQUE PYRAMID

SCALE   DATE

NAME

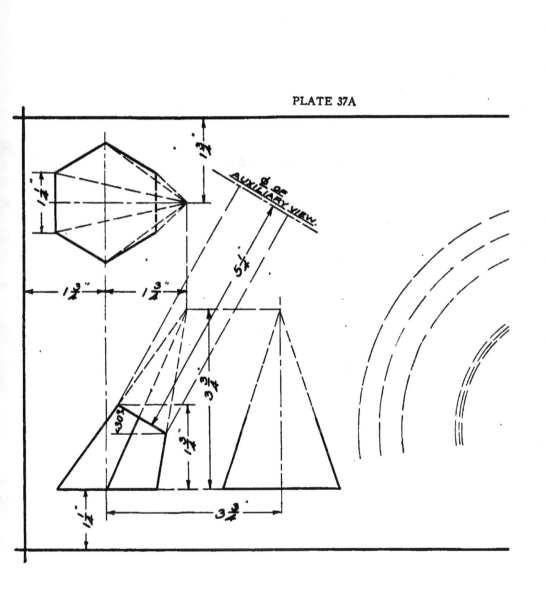

Specification—Plate 38.

Make a drawing, including an auxiliary view, and develop a pattern
Cylinder.   Scale 12″=1′.

Title:—TRUNCATED CYLINDER

SCALE   DATE

NAME

PLATE 38

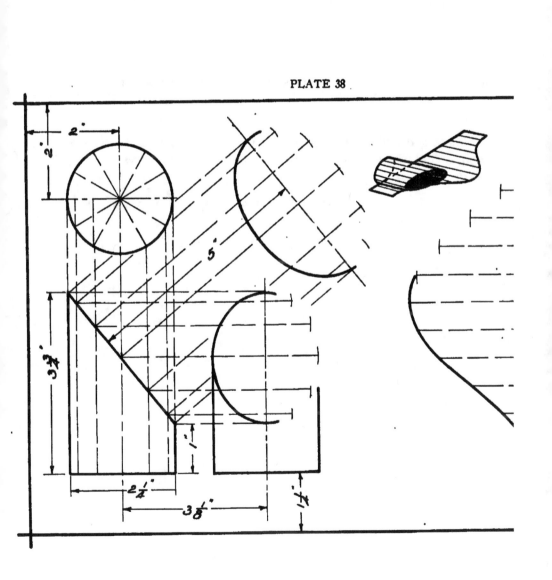

Specification—Plate 38A.

Make a drawing, including an auxiliary view, and develop pattern
Hood, including pattern for handle. The handle is to be made of $\frac{1}{8}''$ x $\frac{3}{4}$
$3''=1'$.

Title:— SHEET-METAL HOOD

        SCALE   DATE

             NAME

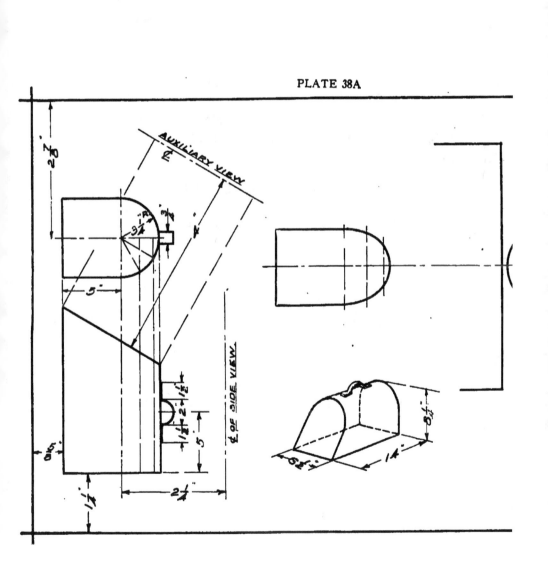

Specification—Plate 39.

Make a drawing, including an auxiliary view, and develop a pattern Cone. Scale 12″—1′.

Title:—   TRUNCATED CONE

SCALE   DATE

NAME

PLATE 39

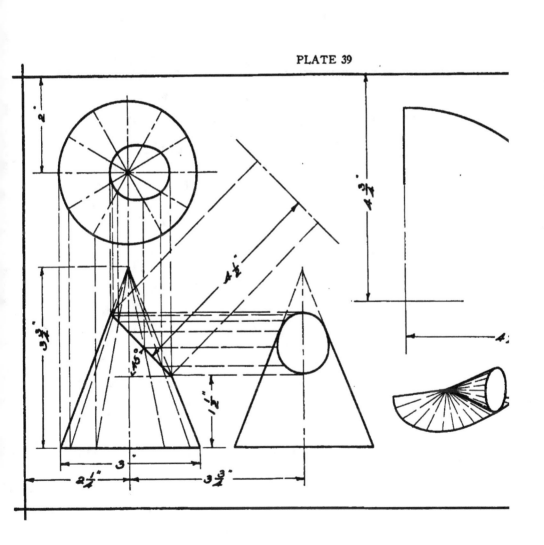

Specification—Plate 39A.

Make a drawing and develop a pattern for the Oblique Cone. Scale

Title:—   OBLIQUE CONE

SCALE   **DATE**

NAME

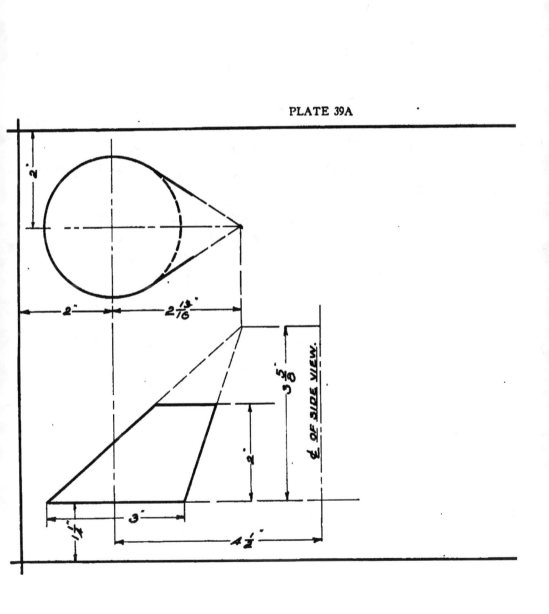

Specification—Plate 40.

Make a drawing of the Intersecting Cylinders, showing the line of in
a pattern for each cylinder.   Scale 12″=1′.

Title :—   INTERSECTING CYLINDERS
                    SCALE   DATE
                            NAME

PLATE 40

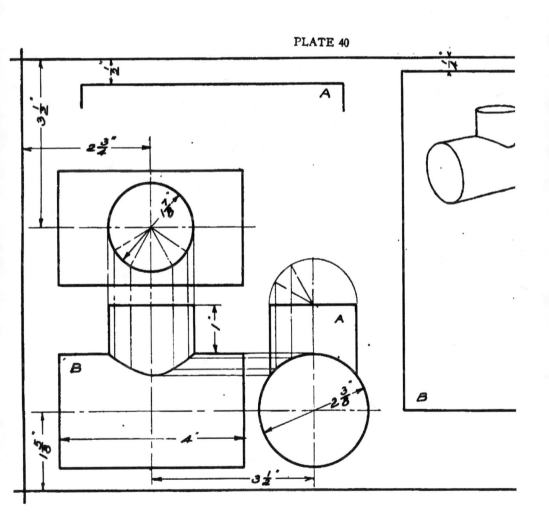

Specification—Plate 40A.

Make a drawing of the Intersecting Cylinders, showing the line of front view and the top view. Develop a pattern for each cylinder. Scal

Title:—   INTERSECTING CYLINDERS

SCALE   DATE

NAME

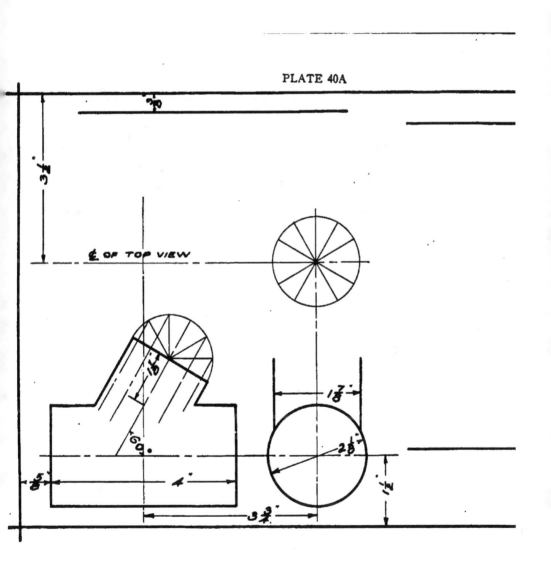

Specification—Plate 41.

Make a drawing of a Cone intersected by a Cylinder as indicated. I intersection in the top view and from it the line of intersection in the fr patterns for the cone and cylinder. Scale $12'' = 1'$.

Title:—   CYLINDER INTERSECTING CONE

SCALE   DATE

NAME

PLATE 41

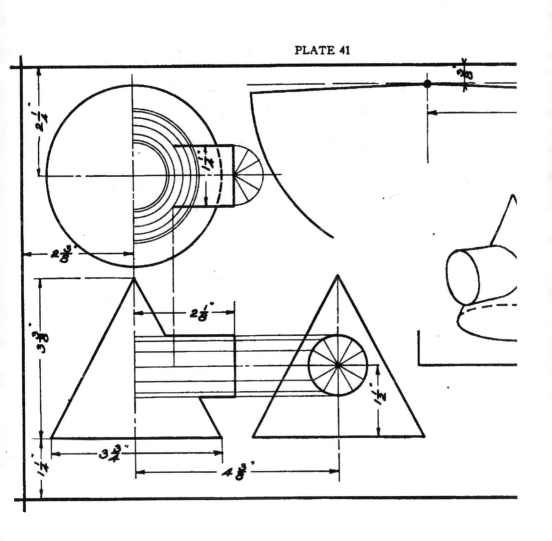

Specification—Plate 41A.

Make a drawing of a Cone intersected by a Cylinder as indicated. intersection in the top view and from it the line of intersection in the f patterns for the cone and cylinder. Scale $12''=1'$.

Title:—   CYLINDER INTERSECTING CONE

SCALE   DATE

NAME

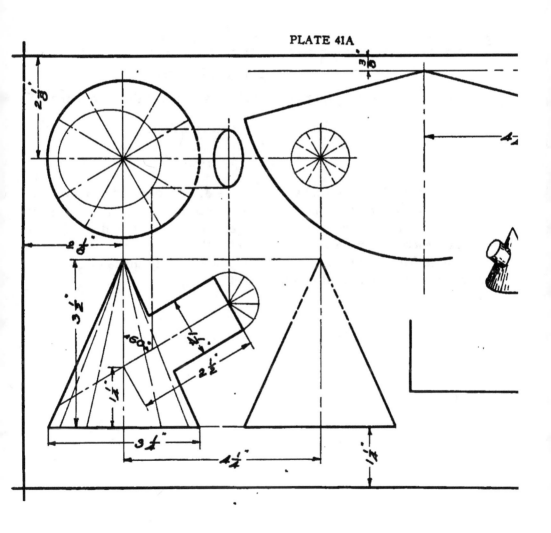

Specification—Plate 42.

Make a working drawing and develop pattern for the Ventilator. S

Title:—   VENTILATOR

SCALE   DATE

NAME

Sheet iron, sheet steel, sheet copper, etc., are formed by the rolling "billet," of heated metal is passed between successive sets of revolving cy as clothes are passed thru a clothes wringer. The metal is reduced in thicl in surface each time it passes thru the rolls, the last set of rolls being so ɛ it to the desired thickness. These rolls are generally made of chilled ca are very heavy.' Metal is also "cold rolled."

PLATE 42

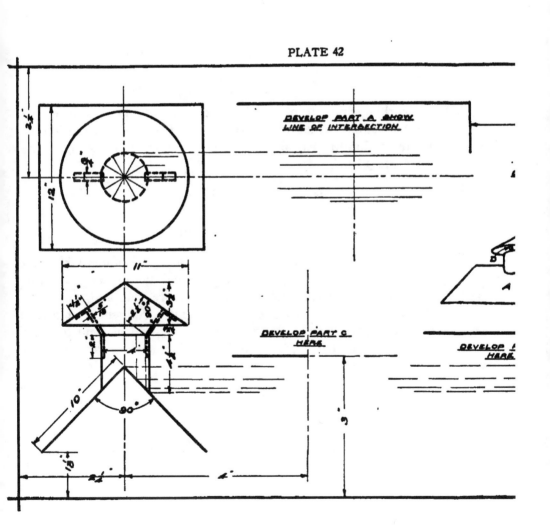

DEVELOP PART A SHOW
LINE OF INTERSECTION

DEVELOP PART C
HERE

DEVELOP
HERE

Specification—Plate 42A.

Make a working drawing of the Pipe Tee, showing the front-view ha
the line of intersection formed by the outside of the pipes on the left sid
the line of intersection formed by the inside of the pipes on the right side
front view. Scale 3″—1′.

Title:—   REDUCING TEE

SCALE   DATE

NAME

Steam, water, oil or gas frequently passes thru pipes under a high pre
must therefore be made carefully so that there will be no leakage at the
pipes and pipe fittings are machined so that they are true and will form
bolted together. To further insure a tight joint, a ring cut from a shee
material is placed between the flanges. This ring, which is cut to the sl
flanges, is called a "gasket." Asbestos, copper, and lead gaskets are a

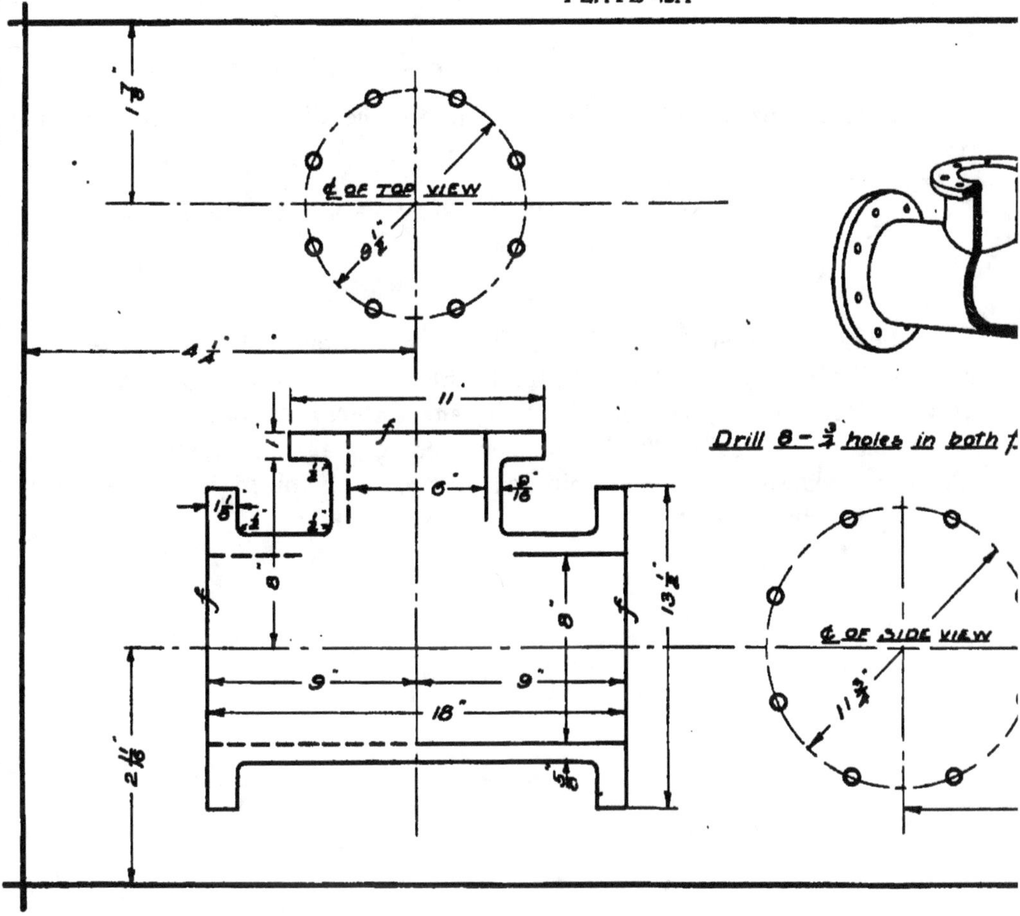

⌀ OF TOP VIEW

Drill 8 - ¾ holes in both f

⌀ OF SIDE VIEW

Specification—Plate 43.

Make drawings and develop patterns for the Pipe Elbow and the Fur
12″=1′.

Title:—   PIPE ELBOW AND FUNNEL
SCALE   DATE
NAME

Sheet iron when exposed to moisture will soon rust. To prevent tl
with a thin coat of some other metal which is not affected by moisture.
is very desirable as the iron gives the sheet thickness and strength, anc
a protection. So-called "tin" which is used to make cans, kitchen utens
sheet iron plated with a very thin coat of tin. Galvanized iron consists (
with a thin coat of zinc.

PLATE 43

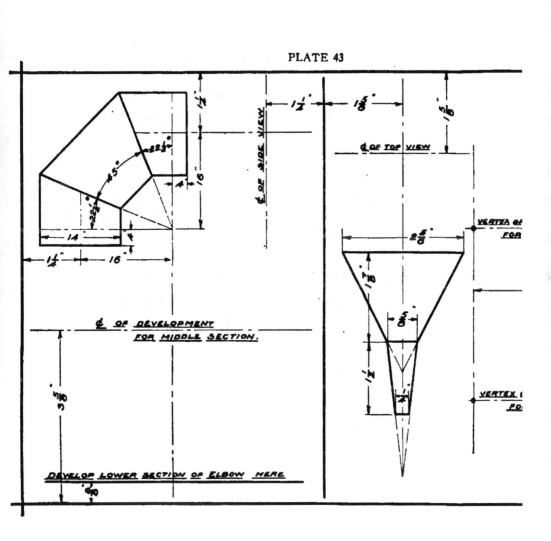

$22\tfrac{1}{2}°$

$45°$

$22\tfrac{1}{2}°$

$\underline{\text{℄ OF SIDE VIEW}}$

$1\tfrac{1}{2}°$

$1\tfrac{5}{8}°$

$1\tfrac{5}{8}°$

$\underline{\text{℄ OF TOP VIEW}}$

$1\tfrac{1}{2}°$

$16°$

$\tfrac{1}{4}$

14

$\tfrac{1}{4}$

$1\tfrac{1}{4}°$

16

VERTEX OF

FOR

$2\tfrac{5}{8}°$

$1\tfrac{7}{16}$

$\tfrac{5}{8}°$

$\underline{\text{℄ OF DEVELOPMENT}}$
$\underline{\text{FOR MIDDLE SECTION.}}$

$1\tfrac{1}{2}°$

VERTEX

FOR

$\tfrac{1}{4}$

$3\tfrac{5}{8}$

$\underline{\text{DEVELOP LOWER SECTION OF ELBOW HERE}}$

$\tfrac{9}{16}$

Specification—Plate 43A.

Make a drawing of the 15° Fork Wrench, using the following values nut.

C — 1$\frac{3}{16}$″,   D = $\frac{11}{16}$″,   H = $\frac{7}{16}$″,   I = 1$\frac{3}{8}$″,   J — 1″,   K — $\frac{7}{8}$″,   L — $\frac{9}{32}$″,

Draw line OA 15° to center line ZY. Draw OB and OE 45° to (ZY as shown. Tangent to circle of C diameter draw jaw faces para half of D on each side of 75° line FG. With x and x′ as centers, dr end of jaw faces to lines OB and OE. With O as center, join the enc inner ends of jaw faces.

The wrench is dropped-forged of steel and is finished all over. §

Title:—   15° FORK WRENCH

SCALE   DATE

NAME

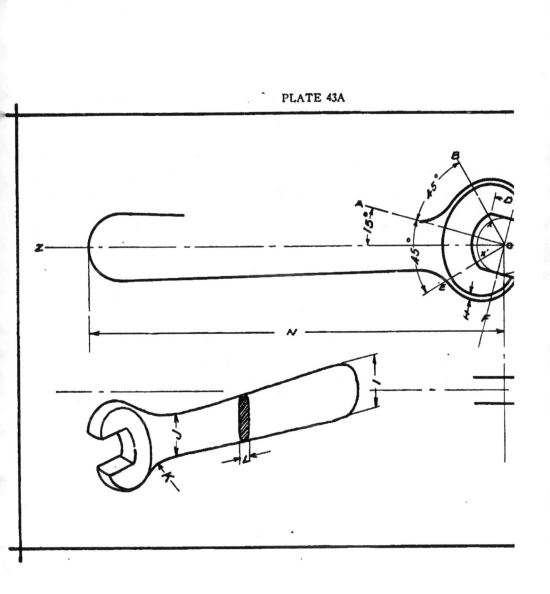

Specification—Plate 44.

Make a drawing of the Steering Column Support, showing the fr
and an auxiliary view as indicated. The lay-out sheet shows the front
as much of the side view as is to be drawn. Complete the auxiliary viev
lines as they do not add any information and would make the drawing
cated. The pads upon which the support rests and the end of the cyli
be faced. Scale 12″=1′.

Title:—   STEERING COLUMN SUPPORT
                SCALE   DATE
                            NAME

AUXILIARY VIEWS.

It will be noticed in the drawing of the Steering Column Support
view shows the shape and construction of the object to a good advantag
are used whenever they make a drawing more clear than can be obtain(

PLATE 44

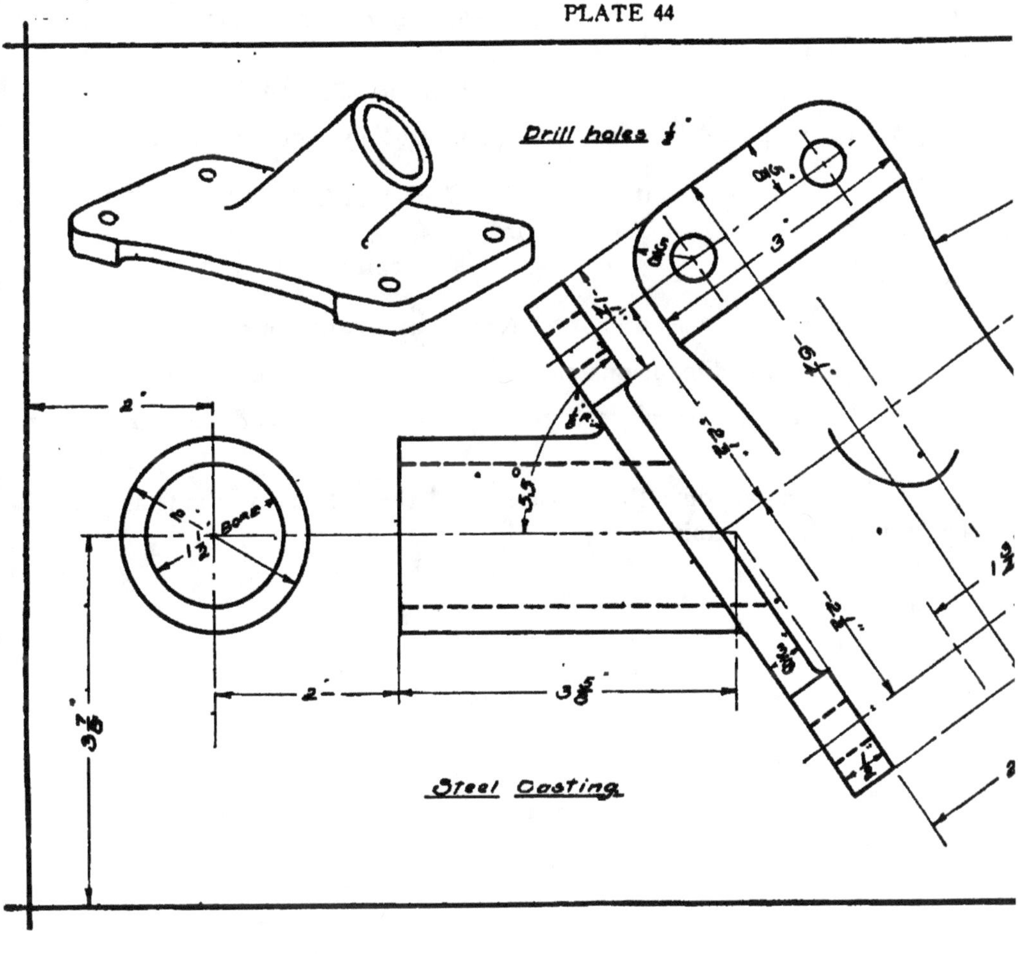

Drill holes ⅜

Steel Casting

Specification—Plate 44A.

Make a working drawing of the Slotted Segment. Scale 12″=1′.

Title:— SLOTTED SEGMENT
SCALE   DATE

NAME

Notes on Dimensions.

1. Important dimensions should not be placed where they may be ov

2. When lines are close together, make arrow-heads so that the wor line they go to.

3. Do not put on all dimensions and then all arrow-heads as you ma arrow-heads by so doing.

CAST BR.

Drill all holes $\frac{7}{16}$

R. OF AUXILIARY VIEW

$1\frac{1}{2}$

$2\frac{1}{4}$

$\frac{5}{8}$

$15°$

$100°$

$135°$

$9\frac{1}{4}$ R.

$\frac{7}{16}$

$9\frac{1}{2}$

$1\frac{1}{4}$

$120°$

$3\frac{1}{4}$

$\frac{9}{16}$

$1\frac{1}{2}$

$3\frac{1}{4}$

$\frac{1}{2}$ R.

$\frac{7}{16}$

$1\frac{3}{4}$

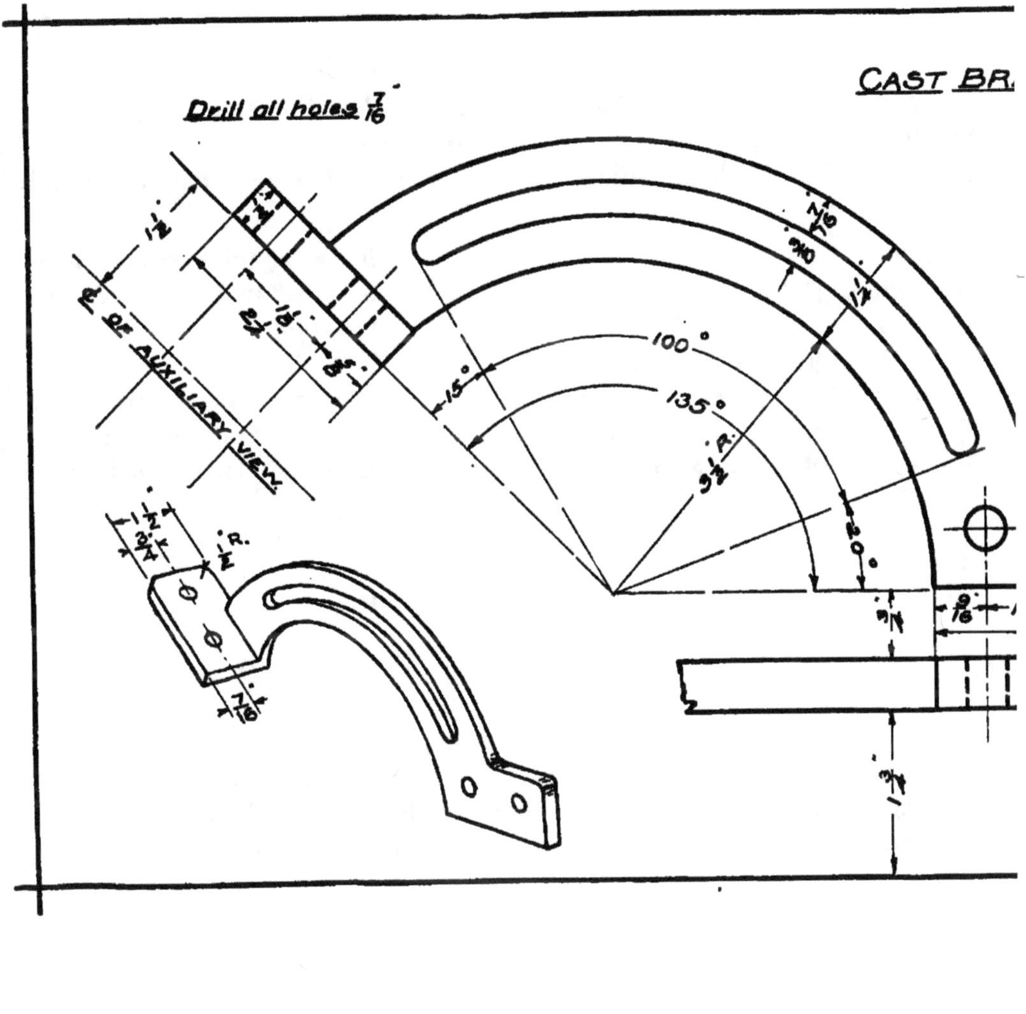

Specification—Plate 45.

Make a drawing of a Cast Iron Pulley 7" in diameter for 3" belt. S
the following values which are taken from formulas and tables by J.
Machinist Hand Book, and are based on the diameter of pulley, width
eter of shaft.

Dia., 7".

W = width, 3½".

S = 1¼" diameter of shaft.

B = 1⅜" width of arm at center.

C = ⅞" width of arm at circumference.

D = ⅜" thickness of arm at center.

E = ¼" thickness of arm at circumference.

F = ¼" thickness of rim at

G = ⅛" thickness of rim at e

O = ½ of C.

Y = E.

I = 3" distance across web.

J = 2¼" diameter of hub.

L = 2½" length of hub.

Title:—   CAST IRON PULLEY

SCALE   DATE

NAME

CONVENTIONAL SECTIONS.

It will be noticed in the drawing of the pulley that the arms are no
side view as would be the case in a true projection from the front view
show clearly the shape of the hub and the rim and is customary in draw
volving spokes, arms, or ribs. The "revolved section" of the arm is a
method of showing the shape of such members or parts. See page 197.

PLATE 45

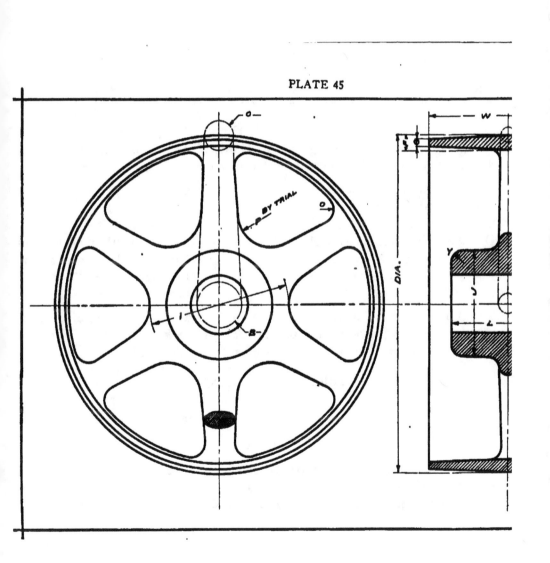

Specification—Plate 45A.

Make a full-size drawing of a 7″ Hand Wheel, using the following derived from formulas and tables based on the diameter of the wheel.

A = 7″ diameter of wheel.

B = $1\frac{3}{8}$″ diameter of rim.

C = ⅝″ offset.

D = 2⅛″ diameter across web.

E = $1\frac{1}{2}$″ thickness of arm at hub.

F = $1\frac{1}{2}$″ thickness of arm at rim.

G = ⅝″.

H = $1\frac{1}{16}$″.

J = $1\frac{1}{16}$″.

K = $1\frac{7}{16}$″.

L = $1\frac{3}{8}$″ width of arm at rim

M = $1\frac{3}{8}$″ width of arm at hub

O = 1.25 x B length of hub.

Diameter of hub = D minu

When reverse curves are to be joined as at x, a short straight line connect them to prevent an apparent kink in the finished curve. This ru when drawing small curves of the same radius. Scale 12″=1′.

Title:—   7″ HAND WHEEL

SCALE   DATE

NAME

CONVENTIONAL SECTIONS.

In a drawing of a pulley or wheel which has an uneven number of sp sectional view should be represented as if there were two arms opposite the hand wheel. The other view should show one of the arms in a verti the horizontal center line.

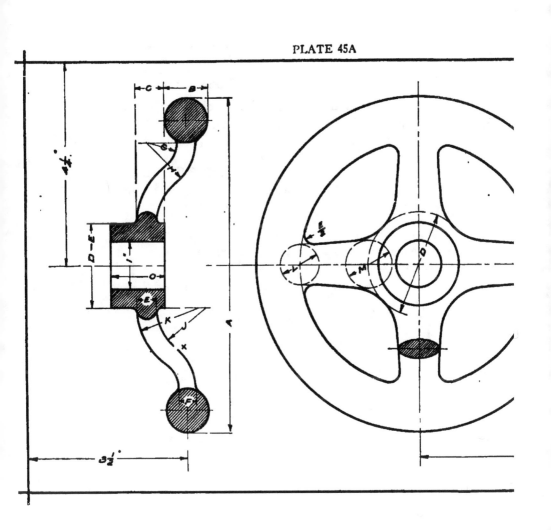

Specification—Plate 46.

Make a drawing of the Library Table, showing the end view in sec
view half in section. Show also, sections on lines V-V, O-O, and X-X. Si
Depth of drawer 19″ over all. All material ¾″ unless otherwise specified.

Title:—   LIBRARY TABLE
SCALE   DATE
NAME

1. Make out a bill of material for the Library Table. The back, sides, and bottom
of basswood. All other pieces are of quarter-sawed white oak.

2. Determine the approximate cost of the material for the table.

PLATE 46

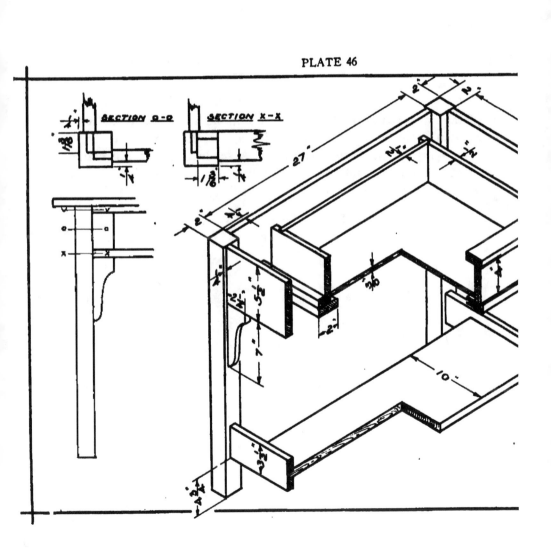

Specification—Plate 46A

Make a drawing of the Bed to a suitable scale. The foot end of the directly in front of the head end with a left-side view in section of the fo side view in section of the head end. A separate drawing is made of the s are fastened to the ends of the bed by means of cast-iron rail fasteners wedge-hook principle.

Title :— SINGLE BED

SCALE   DATE

NAME

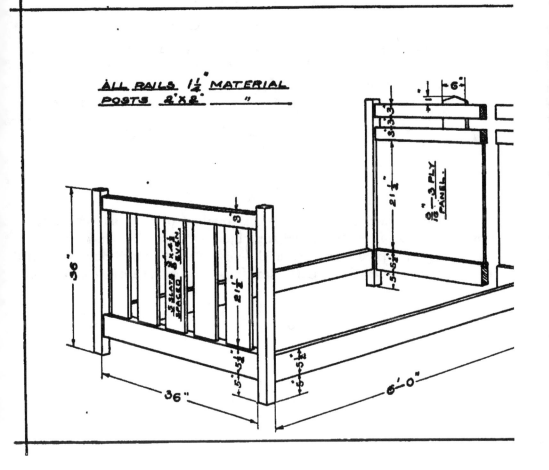

ALL RAILS 1½" MATERIAL
POSTS 2"X2" "

Specification—Plate 47.

Make isometric drawings of a 1⅜″ cube; a circular block ⁵⁄₁₆″ thick resting on a block ⁵⁄₁₆″ thick by 1½″ square; a triangular frame and indicated. In the drawing of the frame and the clutch spider, dimensions on both the working drawing and the isometric drawing. Scale 12″=1′.

Title:— ISOMETRIC DRAWINGS
SCALE   DATE
NAME

ISOMETRIC DRAWING.

Isometric drawing is a mechanical method of pictorial representation represent a complete picture of an object in one view, showing the th height, width, and length. Parallel lines of equal length on the object in the drawing; hence isometric drawings can be dimensioned to a bett perspective drawings.

PLATE 47

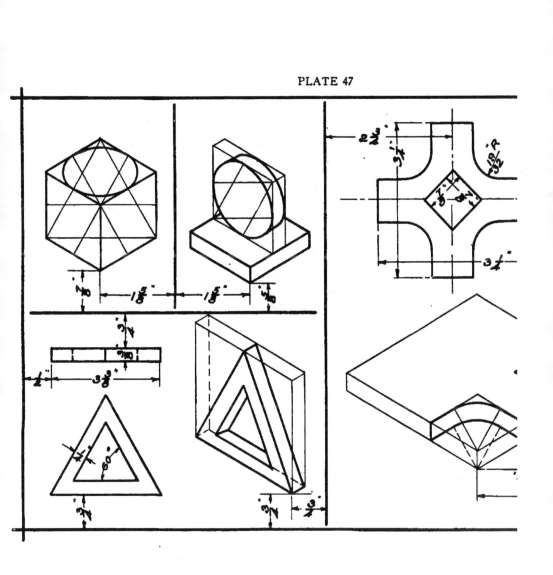

Specification—Plate 47A.

Make an isometric drawing of the Crank as indicated by the workin⸢
the Brace in Plate 7A.   Scale 12″ and 9″ ━ 1′

Title:— ISOMETRIC DRAWINGS
SCALE   DATE
NAME

Dimensions on isometric drawings should be placed so as to read fr⸢
from the *bottom up*. Dimension lines should always be parallel to an isome⸢
cating the diameter of circles it is better to place the dimension outside⸢
the isometric circles.

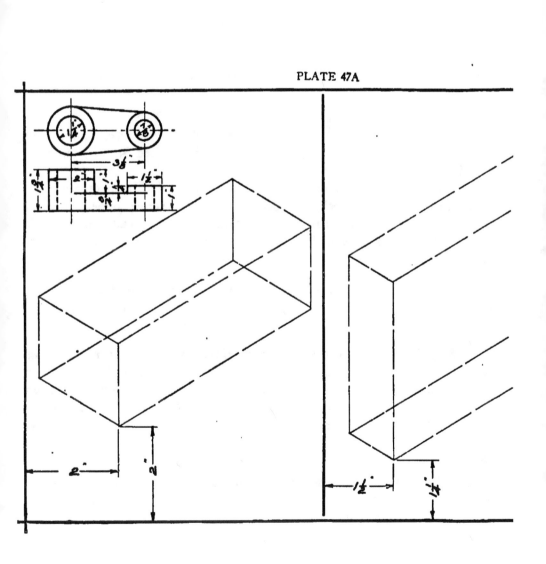

Specification—Plate 48.

1. Draw a helix of two turns having a diameter of 3¼″ and a pitch

2. Draw a profile, or sectional view of the United States Standard ′
as indicated.

3. Draw a conventional representation of screw-threads on a piece of
as indicated. Also a block threaded to fit a 1¼″ screw, upper half in se

4. Draw a conventional representation of screw-threads on a 1″ diai
cated; also a block threaded to fit the rod and an end view of the block a
12″=1′.

Title:—   U. S. STANDARD THREAD

SCALE   DATE

NAME

The helix is the curve of the screw-thread and is the curve used in
representation of a thread. It is the path of a point traced on the surfa
cylinder as the point moves at a uniform rate of speed along a line whi
the axis of the cylinder, and at some regular prescribed proportion of trav
to each revolution of the cylinder. The pitch of the helix is the distance
points in the path of the cylinder measured parallel with the axis of the cy
199 and 200.

In the drawing of screw-threads the actual form of the thread is seld
it involves too much time and work. Conventional forms have been ad
most common being shown in Plate 48.

PLATE 48

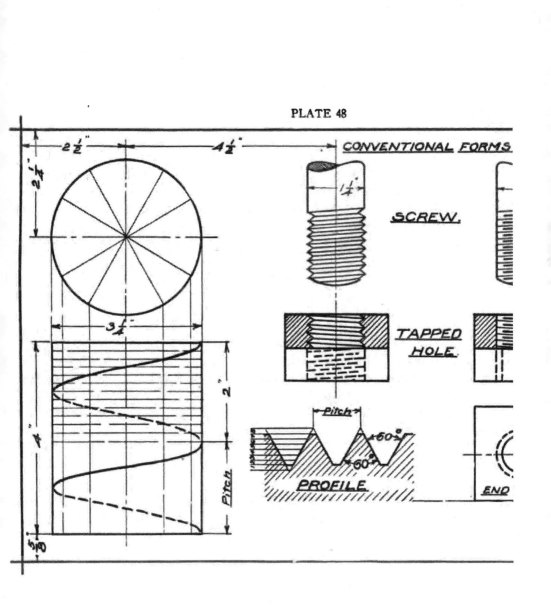

CONVENTIONAL FORMS

SCREW.

TAPPED HOLE.

PROFILE

END

Specification—Plate 48A.

1. Draw a 1". hexagonal bolt and nut, U. S. Standard Thread, usir method of representing the thread as indicated.

2. Draw a 1" square-head bolt and nut, U. S. Standard Thread, usi method of representing thread as indicated. Scale 12"=1'.

For construction of bolt heads and nuts, see page 198.

Title:— MACHINE BOLTS

SCALE   DATE

NAME

There are two classes of bolts, namely: Machine Bolts and Carrie bolts are classed as rough or finished and have square or hexagonal hez of manufacturing rough bolts, rods of iron or steel are cut into pieces according to the length of bolt desired. These pieces are heated at one the jaws of a machine called a bolt header, leaving enough of the hes to form the head. The ram of the machine upsets and forms the heatec shaped head. A thread is cut on the other end of the bolt in a thread threading tool called a die. Finished bolts are turned from hexagonal, so When turned from round bars it is necessary to machine the head to Nuts for the bolts are either punched from heavy sheet-metal or cut froi proper shape. The holes are punched or drilled and are threaded with

PLATE 48A

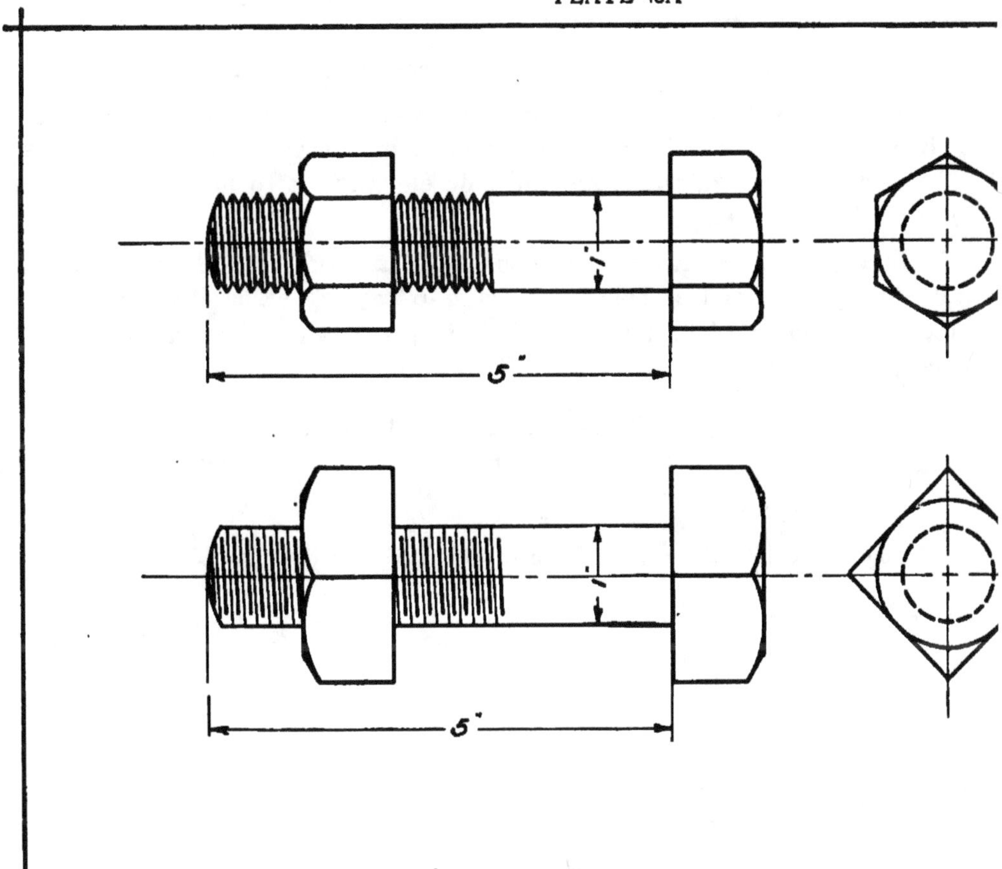

Specification—Plate 49.

Make a working drawing of a 11′ x 16′ Garage. Show the side half in section; front elevation, left-hand half in section; plan view, in se The sectional half in the front and side elevation is to include the concr grade line.

In the front are two hinged doors 3′-6″ x 7′-6″; each door to have a 14″ x 14″ lights and a built-up panel 2′-6″ x 3′-0″. The division bars dow are 2″ wide. The lights are held in place with ½″ x ½″ strips. only for the rear window. Scale ½″=1′.

Title:— AUTOMOBILE GARAGE

SCALE   DATE

NAME

Note:—Use 15″ x 22″ paper. Trim to 14″ x 20½″. Border line 13″ x 19″.

PLATE 49

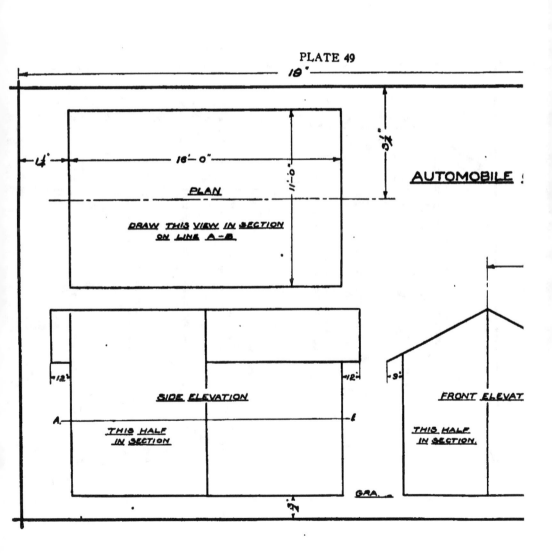

19"

16'-0"

11'-0"

51'

4'

PLAN

DRAW THIS VIEW IN SECTION
ON LINE A-B.

AUTOMOBILE

12"

12"

9"

SIDE ELEVATION

FRONT ELEVAT

A.

B.

THIS HALF
IN SECTION

THIS HALF
IN SECTION.

GRA.

PLATE 49 (Continued)

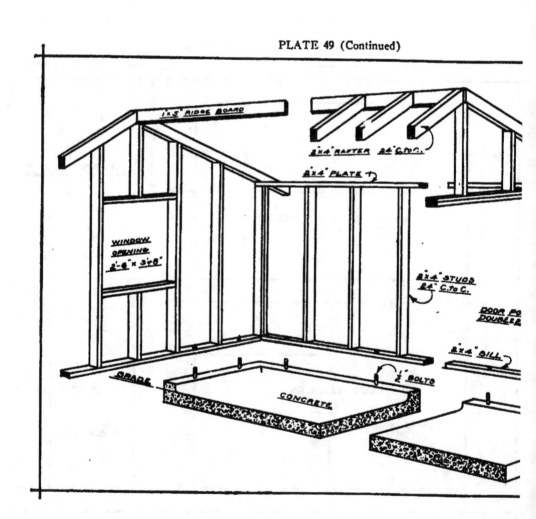

1x5" RIDGE BOARD

2x4" RAFTER 24" C. to C.

2x4" PLATE

WINDOW OPENING 2'-0" x 3'-8"

2x4" STUDS 24" C. to C.

DOOR PO
DOUBLE 2

2x4" SILL

GRADE

½" BOLTS

CONCRETE

PLATE 49 (Continued)

SADDLE BOARD
1"x 4"

SHINGLES

¼ PITCH

¾"x6 DROP SIDING

1" SHEETING

CASING 1⅛"x 3"

6"

6"

VERGE BOARD
1"x 3"

CORNER BOARDS
1⅛"x 5"

6"

JAMB 2"x6"

45° PANEL

GRADE

12"

DOOR

Section through jamb

B

RISE

A

D

SPAN

Specification—Plate 49A.

Make a drawing of the Cornice and Sill to a suitable scale.   Pitch to weather.

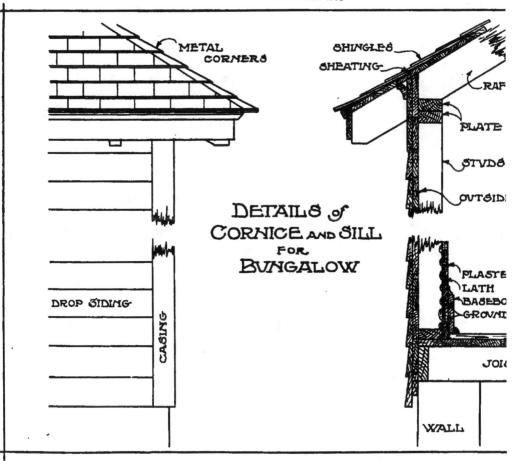

METAL CORNERS

SHINGLES

SHEATING

RAF

PLATE

STVDS

OUTSID

DROP SIDING

CASING

PLASTE

LATH

BASEBO

GROVNI

JOI

WALL

DETAILS of CORNICE and SILL for BUNGALOW

GEOMETRIC PROBLEMS.

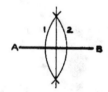

Prob. 1—To bisect a line as AB. With center
radius greater than one-half AB, draw arcs 1 and
of intersection of these arcs draw a line. The line w

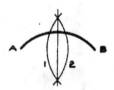

Prob. 2—To bisect the arc of a circle as AB. W
draw intersecting arcs 1 and 2. Draw a line thru
section of these arcs. The line will bisect the arc A1

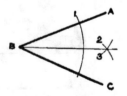

Prob. 3—To bisect an angle as ABC. From B v
as is possible, draw arc 1. From its point of inter:
CB, draw arcs 2 and 3. A line drawn thru B and
2 and 3, will bisect the given angle.

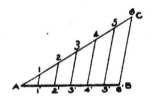

Prob. 4—To divide a line as AB into any number ‹
required number of divisions be six. Draw AC at an'
and lay off six equal spaces of any length. Connect l;
B and draw lines parallel to this line thru the other poin†
in points 1', 2', 3', 4' and 5' which determine the requi

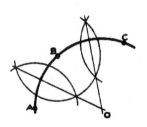

Prob. 5—Given: Three points A, B and C, not in
line. Required: To draw an arc passing thru these poi›
B as centers and any radius greater than one-half AB, de
arcs. With B and C as centers describe similar arcs.
intersections of these arcs. The point of intersection o·
is the center for arc passing thru A, B, and C.

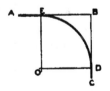

Prob. 6—Given: Lines AB and CB at right angl
Required: To draw an arc of a given radius tangent to †
line DO parallel to AB with distance EO equal to given r
EO parallel to CB with distance DO equal to given ra
center of arc and D and E are the points of tangency.

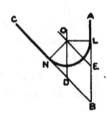

Prob. 7—Given: Lines AB and CB which are
each other. Required: To draw an arc of a given r
lines. Draw line DO parallel to AB with distanc
radius. Draw line EO parallel to CB with distanc
radius. Point O is center of arc and L and N are t

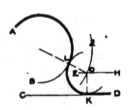

Prob. 8—Given: Any straight line CD and an
To draw an arc of a given radius tangent to line CI
line EH parallel to CD with distance KO equal to
center of arc AB and with radius of arc AB, plus
describe arc 2 passing thru EH. Point O is center
and CD. L and K are the points of tangency.

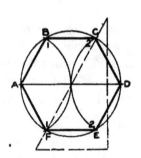

Prob. 9—Given: A circle of any diameter. F
a hexagon within it. Draw a diameter as AD. Wi
and radius equal to that of circle, draw arc 1, 1 and
of intersection A, B, C, D, E, and F to form require

Draftsman's method: Draw horizontal diamet
With 30°, 60° triangle draw line passing thru ce
dotted lines, locating points C and F. In a similar
B and E and connect points with use of 30°, 60° tri
hexagon.

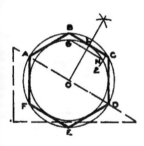

Prob. 10—Given a circle of any diameter. Requi
scribe a hexagon about it. Draw a diameter as BE.
center and radius equal to that of given circle, describe
arc GLN and thru L draw BC parallel to GN. With
radius BO, describe circle BDF. In this circle inscribe

Draftsman's method: With 30°-60° triangle draw c
and CF as illustrated by dotted lines. With same triang
and ED, BC and FE, AF and CD, each tangent to the

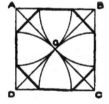

Prob. 11—To draw an octagon within a given square
in square. With A, B, C, and D as centers, strike arcs p
at O and intersecting sides of square. Connect intersect
lines to form octagon.

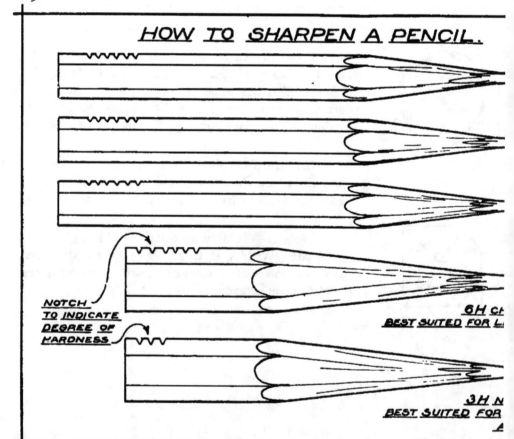

HOW TO SHARPEN A PENCIL.

NOTCH
TO INDICATE
DEGREE OF
HARDNESS

6H C
BEST SUITED FOR L

3H N
BEST SUITED FOR

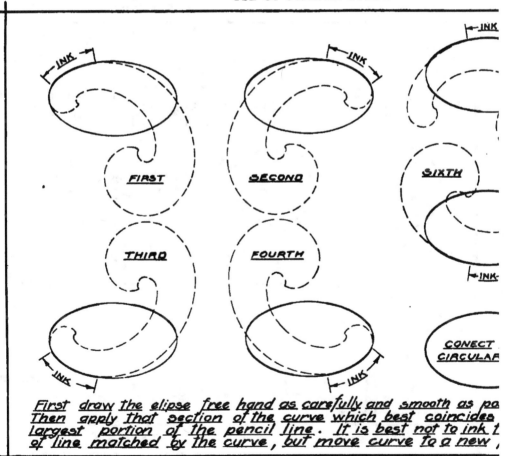

First draw the elipse free hand as carefully and smooth as pa
Then apply that section of the curve which best coincides
largest portion of the pencil line. It is best not to ink t
of line matched by the curve, but move curve to a new

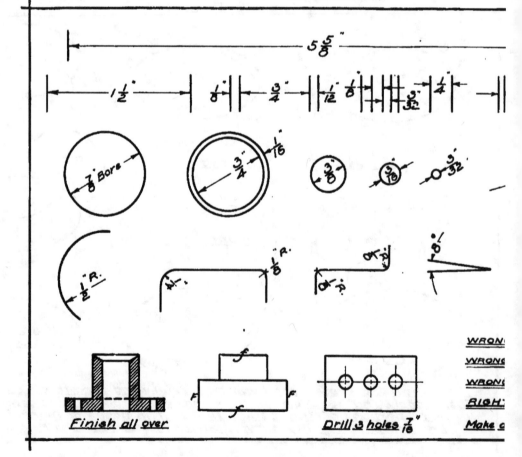

METHODS of INDICATING FINISH.

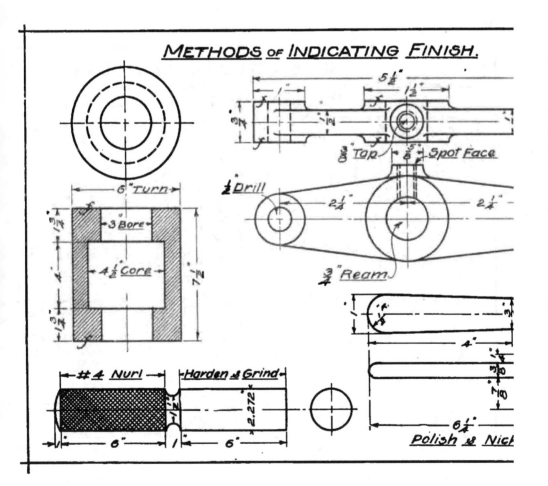

6" Turn

3 Bore

4⅛ Core

7½'

5½"

1½"

⅜" Tap

⅝" Spot Face

½" Drill

2¼"

2¼"

¾" Ream

#4 Nurl

Harden & Grind

2.272

6"

1"

6"

4"

6¼"

Polish & Nick

CONVENTIONAL SECTIONS
AND
REPRESENTATIONS OF MATERIALS

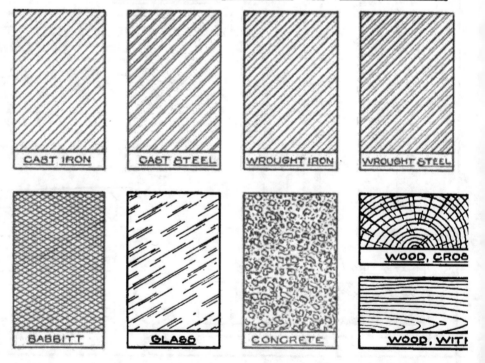

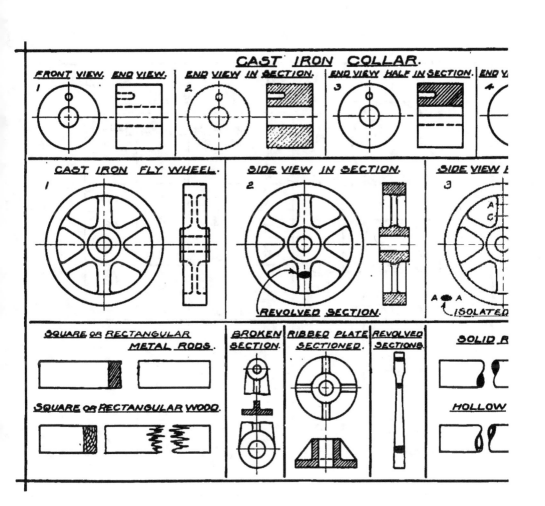

CAST IRON COLLAR.

FRONT VIEW. END VIEW.	END VIEW IN SECTION.	END VIEW HALF IN SECTION.	END V...
1	2	3	4

CAST IRON FLY WHEEL.

	SIDE VIEW IN SECTION.	SIDE VIEW I...
1	2	3

REVOLVED SECTION.

ISOLATED

SQUARE OR RECTANGULAR METAL RODS.

SQUARE OR RECTANGULAR WOOD.

BROKEN SECTION.

RIBBED PLATE SECTIONED.

REVOLVED SECTIONS.

SOLID R...

HOLLOW

U.S. STANDARD BOLTS AND NUTS
(FINISHED)

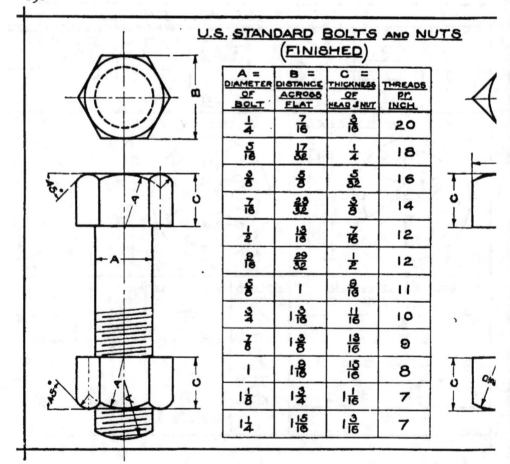

A = DIAMETER OF BOLT	B = DISTANCE ACROSS FLAT	C = THICKNESS OF HEAD & NUT	THREADS pr. INCH
$\frac{1}{4}$	$\frac{7}{16}$	$\frac{3}{16}$	20
$\frac{5}{16}$	$\frac{17}{32}$	$\frac{1}{4}$	18
$\frac{3}{8}$	$\frac{5}{8}$	$\frac{5}{32}$	16
$\frac{7}{16}$	$\frac{23}{32}$	$\frac{3}{8}$	14
$\frac{1}{2}$	$\frac{13}{16}$	$\frac{7}{16}$	12
$\frac{9}{16}$	$\frac{29}{32}$	$\frac{1}{2}$	12
$\frac{5}{8}$	1	$\frac{9}{16}$	11
$\frac{3}{4}$	$1\frac{3}{16}$	$\frac{11}{16}$	10
$\frac{7}{8}$	$1\frac{3}{8}$	$\frac{13}{16}$	9
1	$1\frac{9}{16}$	$\frac{15}{16}$	8
$1\frac{1}{8}$	$1\frac{3}{4}$	$1\frac{1}{16}$	7
$1\frac{1}{4}$	$1\frac{15}{16}$	$1\frac{3}{16}$	7

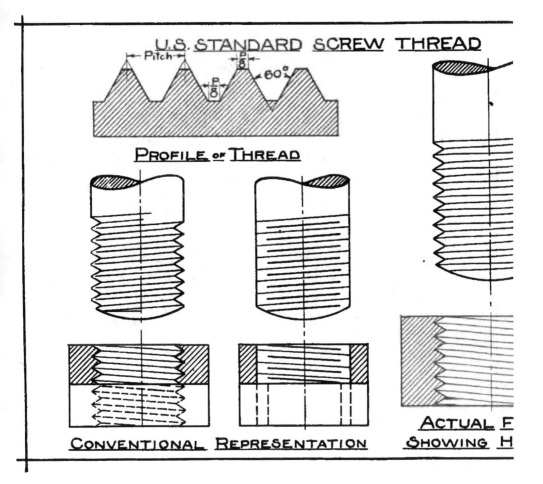

U.S. STANDARD SCREW THREAD

Pitch | $\frac{P}{8}$ | $\frac{P}{8}$ | 60°

PROFILE OF THREAD

CONVENTIONAL REPRESENTATION

ACTUAL F
SHOWING H

202

CABINET DRAWING ILLUSTRATED

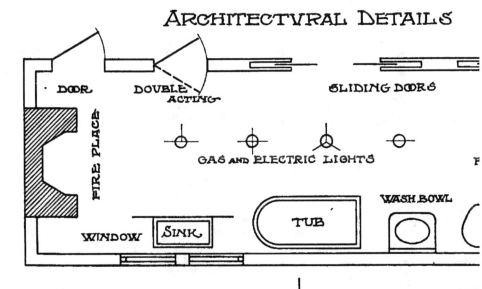

ARCHITECTVRAL DETAILS

DOOR DOUBLE ACTING SLIDING DOORS

FIRE PLACE

GAS AND ELECTRIC LIGHTS

WINDOW SINK TUB WASH BOWL

ABCDEFGHIJKLMNO | ABCDEFGHIJ
PQRSTUVWXYZ & | PQRSTUVW
abcdefghijklmnopqrstuvw

STRUCTURAL STEEL FORM.

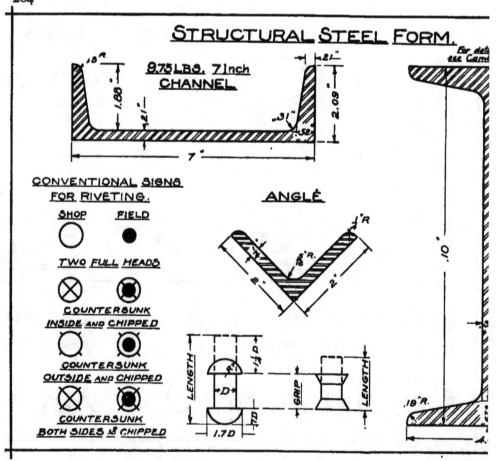

9.75 LBS. 7 Inch CHANNEL

1.88" .21" 2.09" 7" .31" .18"R.

CONVENTIONAL SIGNS FOR RIVETING.

SHOP   FIELD

TWO FULL HEADS

COUNTERSUNK INSIDE AND CHIPPED

COUNTERSUNK OUTSIDE AND CHIPPED

COUNTERSUNK BOTH SIDES & CHIPPED

ANGLE

For details see Cam

LENGTH   1.7 D   .7 D   D   .7 D

GRIP   LENGTH

.18"R.

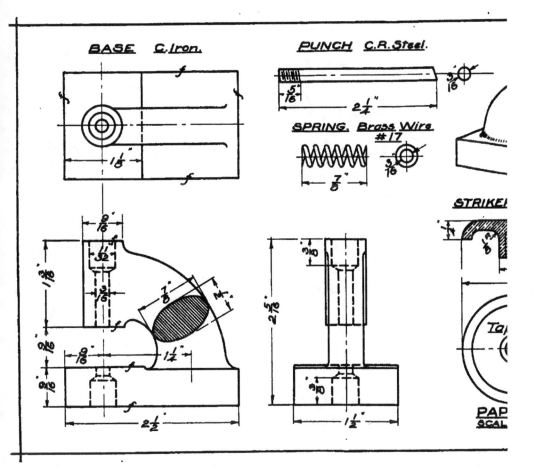

BASE C.Iron.

PUNCH C.R.Steel.

SPRING. Brass Wire
#17

STRIKER

Tap

PAP
SCAL

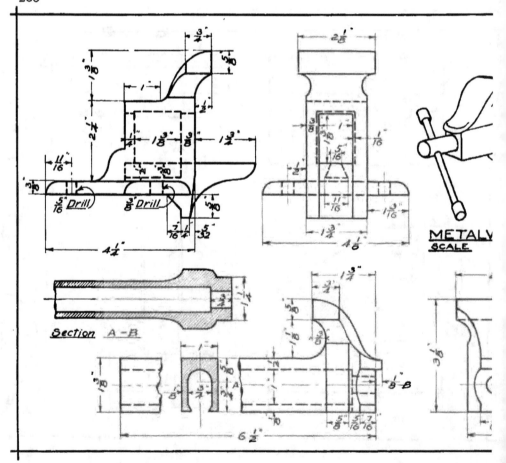

Section A - B

METALV
SCALE

SCREW Mch. Steel. Finish all over.

Pitch $\frac{1}{4}$

$6"$

$\frac{21}{64}$

$\frac{1}{4}$

$\frac{1}{4}$

$\frac{3}{64}"$

$7\frac{1}{8}"$

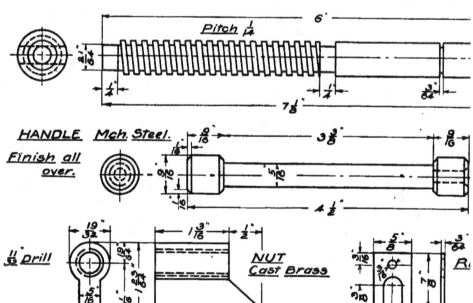

HANDLE Mch. Steel.
Finish all over.

$\frac{9}{16}$

$\frac{1}{16}$

$\frac{9}{16}$

$3\frac{3}{8}"$

$\frac{9}{16}$

$\frac{5}{16}$

$\frac{1}{16}$

$4\frac{1}{2}"$

$\frac{19}{32}"$

$\frac{11}{32}"$ Drill

$\frac{15}{64}$

$\frac{23}{64}$

$1\frac{5}{16}$

$\frac{5}{16}$

$1\frac{3}{16}"$

$\frac{1}{2}"$

NUT
Cast Brass

$\frac{3}{16}$

$\frac{11}{16}"$

$1\frac{11}{16}"$

$\frac{5}{8}"$

$\frac{3}{16}$

$\frac{3}{8}"$

$\frac{7}{8}"$

$\frac{21}{64}$

$\frac{3}{64}"$

R

VIS

SCALE

208

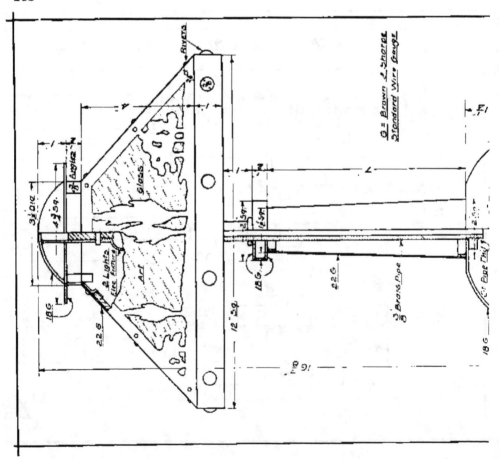

G = Brown & Sharpe
Standard Wire Gauge

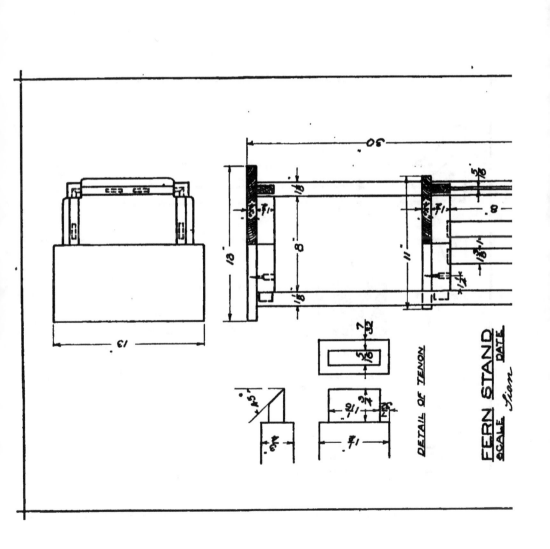

FERN STAND

SCALE _from_

DATE

DETAIL OF TENON

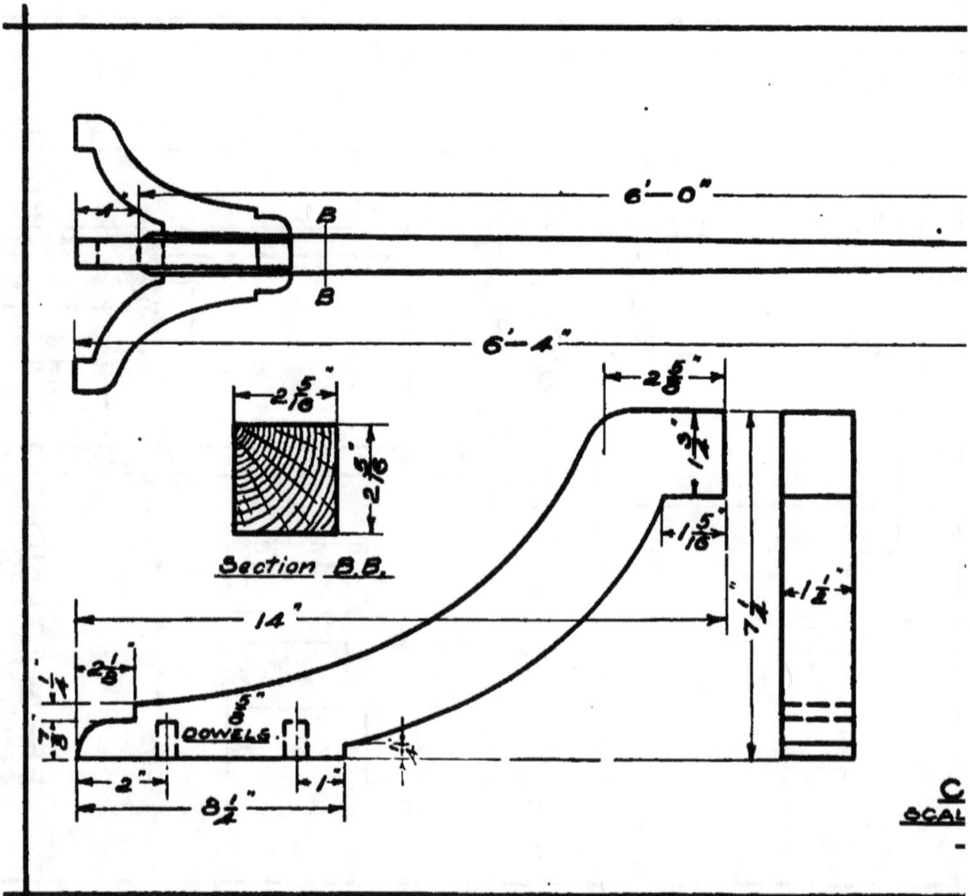

6'-0"

6'-4"

B
B

4

2 5/16"

2 5/16"

Section B.B.

14"

2 5/8"

3/4"

1 5/16"

7 1/4"

1 1/2"

2 1/8"

5/8"

7/8"

DOWELS

2"

1"

8 1/2"

C
SCAL

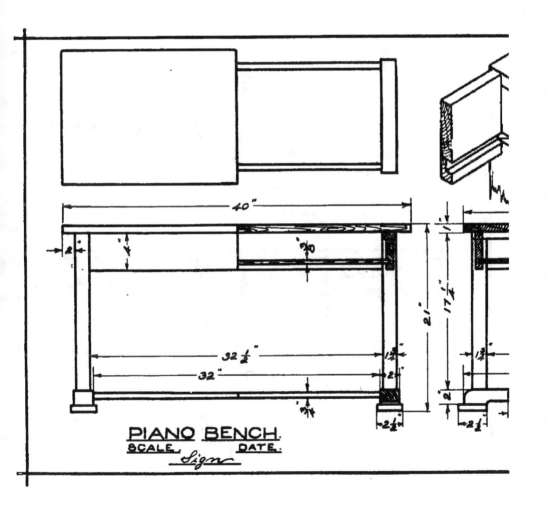

PIANO BENCH.
SCALE. DATE.
Sign

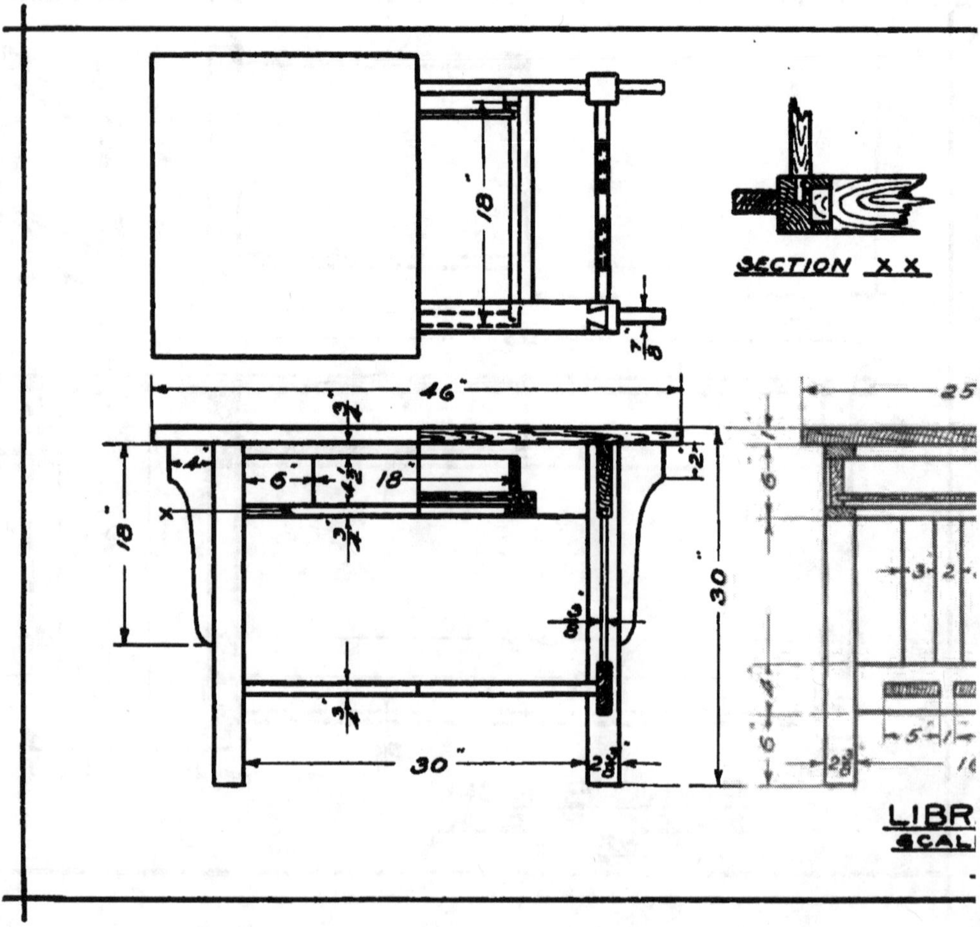

SECTION X X

18"

7/8"

46"

3/4"

4"

6"

18"

3/8"

18"

30"

3"

30"

2"

25

6"

6"

3" 2"

6"

2 3/8"

5" 1

16

LIBR

SCAL

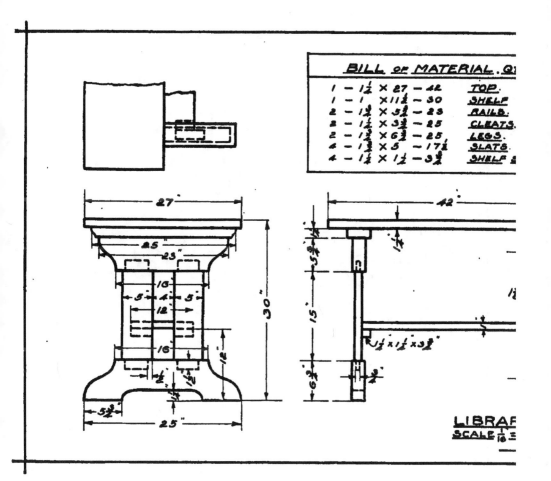

LIBRAR

SCALE $\frac{1}{16}$ =

TURNED
PEDESTAL SCALE DATE

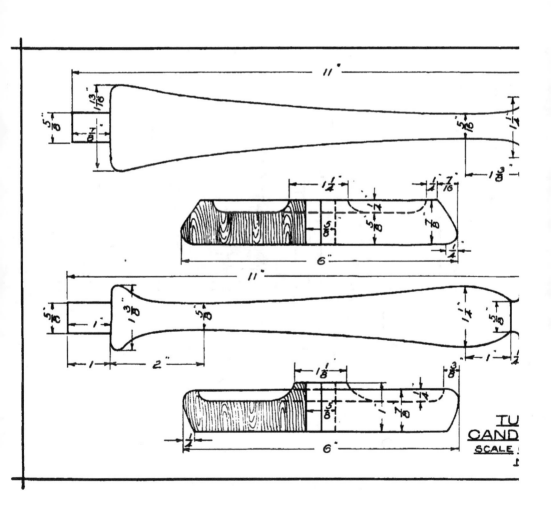

TU
CAND

SCALE

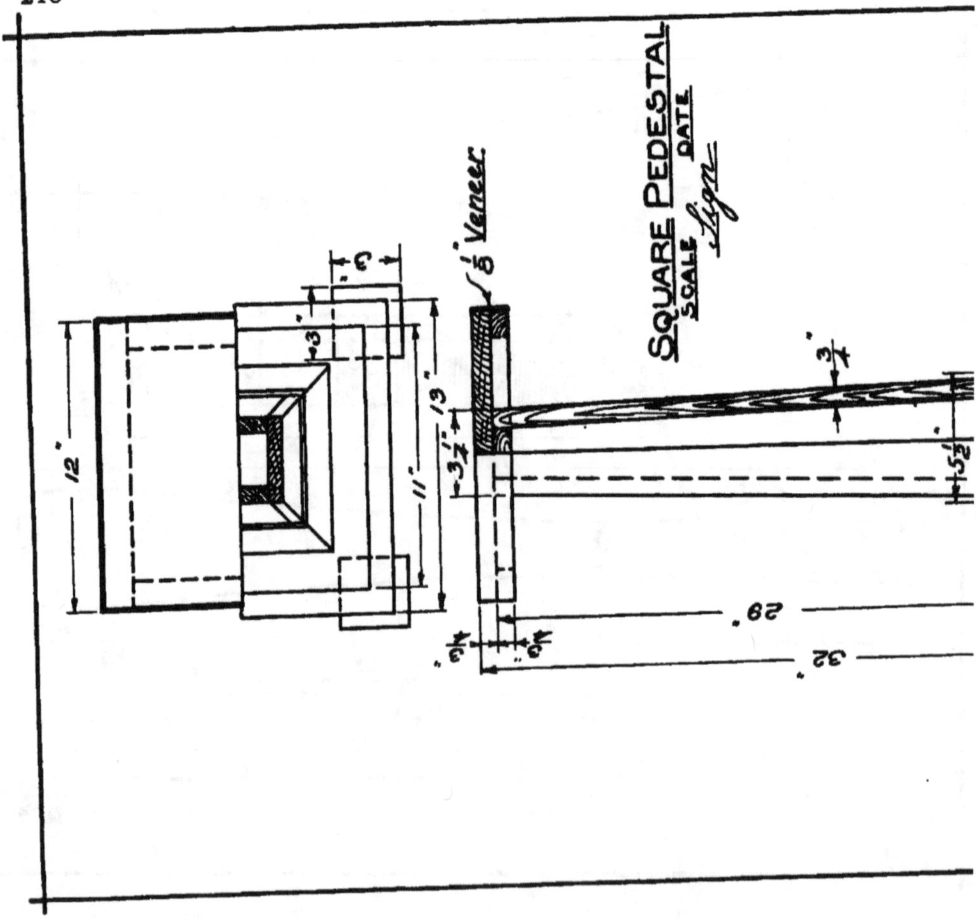

SQUARE PEDESTAL
SCALE   DATE
Sign

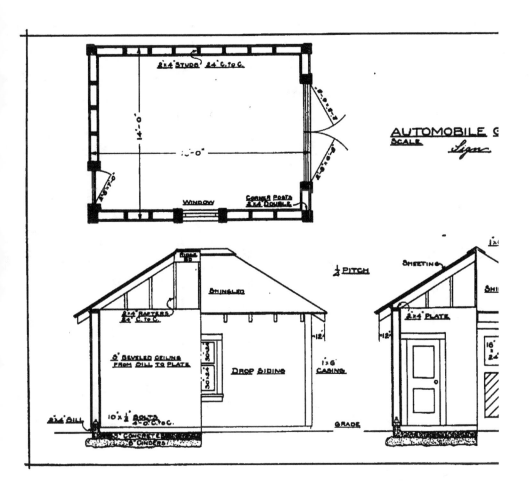

AUTOMOBILE G
SCALE

2×4 STUDS 24" C. to C.

14'-0"

16'-0"

CORNER POSTS
2×4 DOUBLE

WINDOW

RIDGE

½ PITCH

SHEETING

SHINGLED

2×4 RAFTERS
24" C. to C.

2×4 PLATE

SHI

6" BEVELED CEILING
FROM SILL TO PLATE

DROP SIDING

1×6
CASING

2×4 SILL

10"×½" BOLTS
4'-0" C. to C.

GRADE

CONCRETE
CINDERS

WINDOW DETAILS

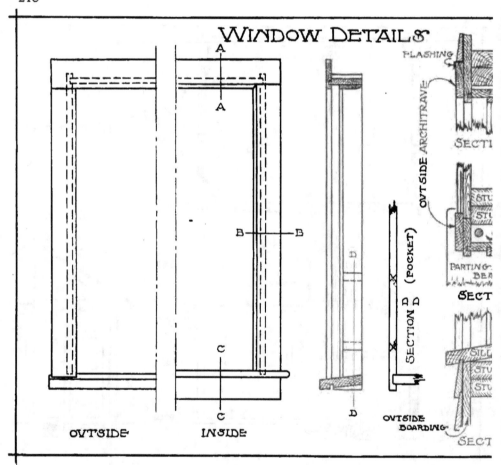

OUTSIDE ARCHITRAVE

FLASHING

SECTI

STU
STU

PARTING
BEA

SECT

SILL
STU
STU

SECT

A
A

B — B

C
C

SECTION D (POCKET)

D

D

OUTSIDE
BOARDING

OUTSIDE   INSIDE

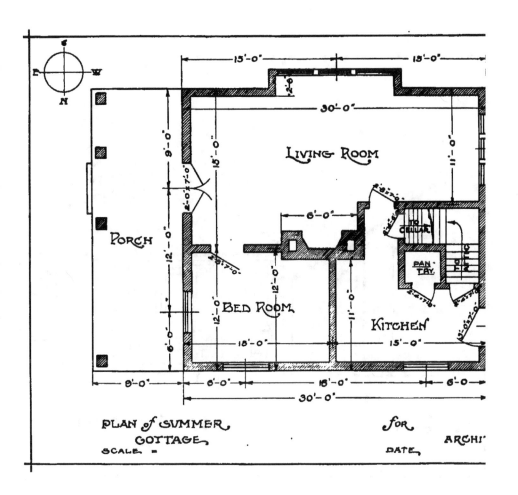

PLAN of SUMMER
COTTAGE
SCALE =

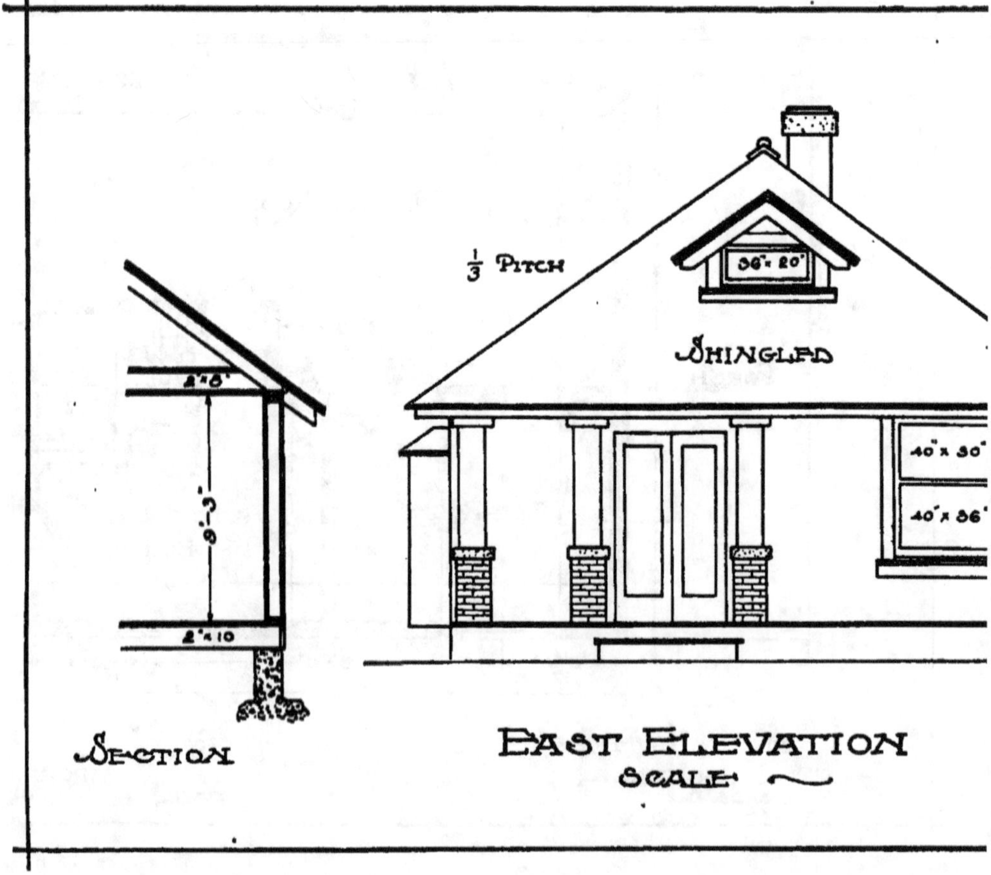

$\frac{1}{3}$ PITCH

36" × 20"

SHINGLED

2'6"

9'-3"

40" × 30"

40" × 36"

2'×10'

SECTION

EAST ELEVATION

SCALE

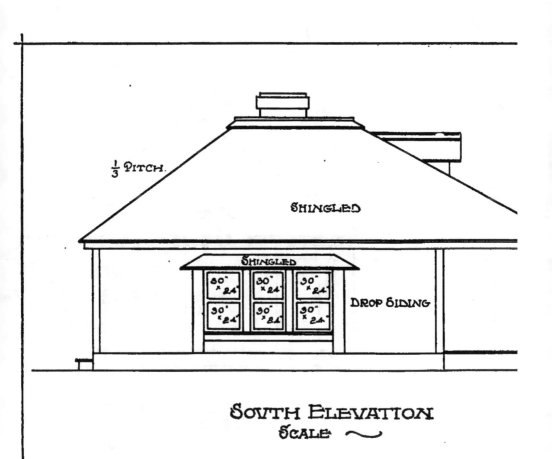

$\frac{1}{3}$ Pitch.

Shingled

Shingled

Drop Siding

30"
× 24"

30"
× 24"

30"
× 24"

30"
× 24"

30"
× 24"

30"
× 24"

SOUTH ELEVATION
Scale

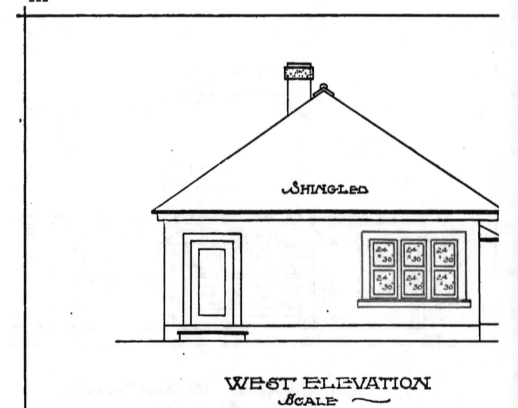

WEST ELEVATION

Scale

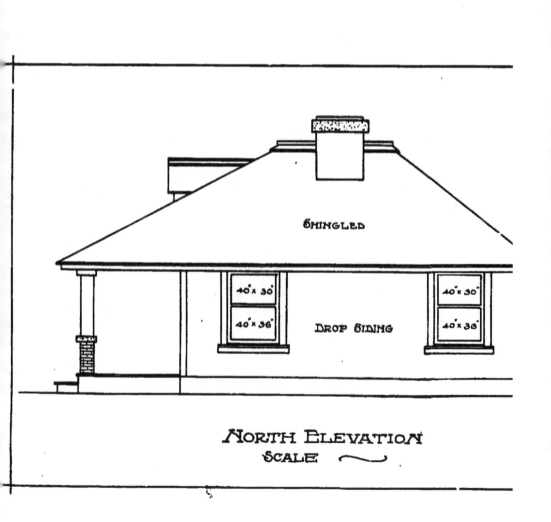

SHINGLED

40"x 30"

40"x 36"

DROP SIDING

40"x 30"

40"x 36"

NORTH ELEVATION

SCALE